Student Solutions Manual

Charles W. Haines
Rochester Institute of Technology

to accompany

Elementary Differential Equations
Sixth Edition

and

Elementary Differential Equations and Boundary Value Problems
Sixth Edition

William E. Boyce
Richard C. DiPrima
Rensselaer Polytechnic Institute

John Wiley & Sons, Inc.
New York · Chichester · Brisbane · Toronto · Singapore

ISBN 0-471-13582-8

Printed in the United States of America

10 9 8 7 6 5 4 3 2 1

Printed and bound by Malloy Lithographing, Inc.

PREFACE

This supplement has been prepared for use in conjunction with the sixth editions of ELEMENTARY DIFFERNTIAL EQUATIONS AND BOUNDARY VALUE PROBLEMS and ELEMENTARY DIFFERENTIAL EQUATIONS, both by W.E. Boyce and R.C. DiPrima. The supplement contains a sampling of the problems from each section of the text. In most cases the complete details in determining the solutions are given while in the remainder of the problems helpful hints are provided. The problems chosen in each section represent, wherever possible, the variety of applications and types of examples that are covered in the written material of the text, thereby providing the student a complete set of examples from which to learn.

Students should be aware that following these solutions is very different from designing and constructing one's own solution. Using this supplemental resource appropriately for learning differential equations is outlined as follows:

   1. Make an honest attempt to solve the problem without using the guide.
   '2. If needed, glance at the beginning of the solution in the guide and then try again to generate the complete solution. Continue using the guide for hints when you reach an impasse.
   3. Compare your final solution with the one provided to see whether yours is more or less efficient than the guide, since there is frequently more than one correct way to solve a problem.
   4. Ask yourself why that particular problem was assigned.

The use of a symbolic computational software package, in many cases, would greatly simplify finding the solution to a given problem, but the details given in this solutions manual are important for the student's understanding of the underlying mathematical principles and applications. In other cases, these software packages are essential for completing the given problem, as the calculations would be overwhelming using analytical techniques. In these cases, some steps or hints are given and then reference made to the use of an appropriate software package.

In order to simplify the text, the following abbreviations have been used:

D.E.	differential equation(s)
O.D.E.	ordinary differential equation(s)
P.D.E.	partial differential equation(s)
I.C.	initial condition(s)
I.V.P.	initial value problem(s)
B.C.	boundary condition(s)
B.V.P.	boundary value problem(s)

I wish to express my appreciation to Mrs. Susan A. Hickey and Ms. Tracey E. Brown for their excellent typing and proofreading of all stages of the manuscript. Dr. Josef S. Torok has also provided invaluable assistance with the preparation of all the figures as well as assistance with many of the solutions involving the use of symbolic computational software.

Charles W. Haines
Professor of Mathematics and
    Mechanical Engineering
Rochester Institute of Technology
Rochester, New York
June, 1996

CONTENTS

CHAPTER 1

<u>Section 1.1, Page 10</u>

2.   The D.E. is second order since there is a second derivative of y appearing in the equation. The equation is nonlinear due to the y^2 term (as well as due to the y^2 term multiplying the y″ term).

6.   This is a third order D.E. since the highest derivative is y‴ and it is linear since y and all its derivatives appear to the first power only. The terms t^2 and $\cos^2 t$ do not affect the linearity of the D.E.

8.   For $y_1(t) = e^{-3t}$ we have $y_1'(t) = -3e^{-3t}$ and $y_1''(t) = 9e^{-3t}$. Substitution of these into the D.E. yields
$$9e^{-3t} + 2(-3e^{-3t}) - 3(e^{-3t}) = (9-6-3)e^{-3t} = 0.$$

14.   Recall that if $u(t) = \int_0^t f(s)\,ds$, then $u'(t) = f(t)$.

16.   Differentiating e^{rt} twice and substituting into the D.E. yields $r^2 e^{rt} - e^{rt} = (r^2-1)e^{rt}$. If $y = e^{rt}$ is to be a solution of the D.E. then the last quantity must be zero for all t. Thus $r^2-1 = 0$ since e^{rt} is never zero.

19.   Differentiating t^r twice and substituting into the D.E. yields $t^2[r(r-1)t^{r-2}] + 4t[rt^{r-1}] + 2t^r = [r^2+3r+2]t^r$. If $y = t^r$ is to be a solution of the D.E., then the last term must be zero for all t and thus $r^2 + 3r + 2 = 0$.

22.   The D.E. is second order since there are second partial derivatives of u(x,y) appearing. The D.E. is nonlinear due to the product of u(x,y) times u (or u).
$$_x_y$$

26.   Since $\dfrac{\partial u_1}{\partial t} = -\alpha^2 e^{-\alpha^2 t}\sin x$ and $\dfrac{\partial^2 u_1}{\partial x^2} = -e^{-\alpha^2 t}\sin x$ we have $\alpha^2[-e^{-\alpha^2 t}\sin x] = -\alpha^2 e^{-\alpha^2 t}\sin x$, which is true for all t and x.

29. Observing the direction field,
     we see that for y>-1/2 we have
     y'<0, so the solution is
     decreasing here. Likewise,
     for y<-1/2 we have y'>0 and
     thus y(t) is increasing here.
     Since the slopes get closer to
     zero as y gets closer to -1/2,
     we conclude that y→-1/2 as t→∞.

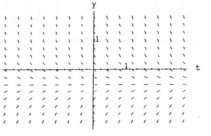

33.

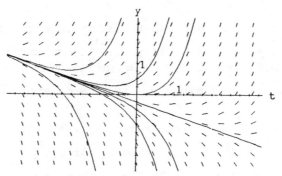

34.

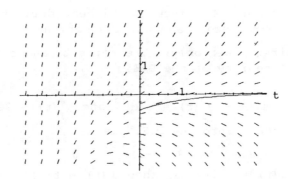

38.

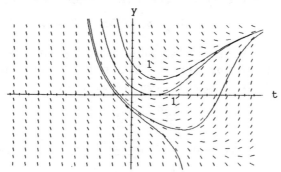

CHAPTER 2

1. $\mu(t) = \exp(\int 3dt) = e^{3t}$. Thus $e^{3t}(y'+3y) = e^{3t}(t+e^{-2t})$ or $\frac{d}{dt}(ye^{3t}) = te^{3t} + e^t$. Integration of both sides yields $ye^{3t} = \frac{1}{3}te^{3t} - \frac{1}{9}e^{3t} + e^t + c$, and division by e^{3t} gives the general solution. Note that $\int te^{3t}dt$ is evaluated by integration by parts, with $u = t$ and $dv = e^{3t}dt$.

2. $\mu(t) = e^{-2t}$.   3. $\mu(t) = e^t$.

4. $\mu(t) = \exp(\int \frac{dt}{t}) = e^{\ln t} = t$, so $(ty)' = 3t\cos 2t$, and integration by parts yields the general solution.

6. The equation must be divided by t so that it is in the form of Eq.(2): $y' + (2/t)y = (\sin t)/t$. Thus $\mu(t) = \exp(\int \frac{2dt}{t} = t^2$, and $(t^2y)' = t\sin t$. Integration then yields $t^2y = -t\cos t + \sin t + c$.

7. $\mu(t) = e^{t^2}$.   8. $\mu(t) = \exp(\int \frac{4tdt}{1+t^2}) = (1+t^2)^2$.

11. $\mu(t) = e^t$ so $(e^ty)' = 5e^t\sin 2t$. To integrate the right side you can integrate by parts (twice), use an integral table, or use a symbolic computational software program to find $e^ty = 5(\sin 2t - 2\cos 2t)/5 + c$.

13. $\mu(t) = e^{-t}$ and $y = 2(t-1)e^{2t} + ce^t$. To find the value for c, set $t = 0$ in y and equate to 1, the initial value of y. Thus $-2+c = 1$ and $c = 3$, which yields the solution of the given initial value problem.

15. $\mu(t) = \exp(\int \frac{2dt}{t}) = t^2$ and $y = t^2/4 - t/3 + 1/2 + c/t^2$. Setting $t = 1$ and $y = 1/2$ we have $c = 1/12$.

18. $\mu(t) = t^2$. Thus $(t^2y)' = t\sin t$ and $t^2y = -t\cos t + \sin t + c$. Setting $t = \pi/2$ and $y = 1$ yields $c = \pi^2/4 - 1$.

20. $\mu(t) = te^t$.

21b. $\mu(t) = e^{-t/2}$ so $(e^{-t/2}y)' = 2e^{-t/2}\cos t$.   Integrating (see comments in #11) and dividing by $e^{-t/2}$ yields

$y(t) = -\dfrac{4}{5}\cos t + \dfrac{8}{5}\sin t + ce^{t/2}$.   Thus $y(0) = -\dfrac{4}{5} + c = a$,

or $c = a + \dfrac{4}{5}$ and $y(t) = -\dfrac{4}{5}\cos t + \dfrac{8}{5}\sin t + (a + \dfrac{4}{5})e^{t/2}$.

If $(a + \dfrac{4}{5}) = 0$, then the solution is oscillatory for all

t, while if $(a + \dfrac{4}{5}) \neq 0$, the solution is unbounded as

$t \to \infty$.   Thus $a_0 = -\dfrac{4}{5}$.

21a.

24a.

24b. $\mu(t) = \exp\displaystyle\int \dfrac{2dt}{t} = t^2$, so $(t^2 y)' = \sin t$ and

$y(t) = \dfrac{-\cos t}{t^2} + \dfrac{c}{t^2}$.   Setting $t = -\dfrac{\pi}{2}$ yields

$\dfrac{4c}{\pi^2} = a$ or $c = \dfrac{a\pi^2}{4}$ and hence $y(t) = \dfrac{a\pi^2/4 - \cos t}{t^2}$, which

is unbounded as $t \to 0$ unless $a\pi^2/4 = 1$ or $a_0 = 4/\pi^2$.

24c. For $a = 4/\pi^2$ $y(t) = \dfrac{1 - \cos t}{t^2}$.   To find the limit as

$t \to 0$ L'Hopital's Rule must be used:

$\lim\limits_{t\to 0} y(t) = \lim\limits_{t\to 0}\dfrac{\sin t}{2t} = \lim\limits_{t\to 0}\dfrac{\cos t}{2} = \dfrac{1}{2}$.

28. The D.E. as given is nonlinear. However, if we think of y as the independent variable and t as the dependent variable then the D.E. can be written as $\dfrac{dt}{dy} = e^y - t$ or

$\dfrac{dt}{dy} + t = e^y$. The integrating factor is then $\mu(y) = e^y$ and the I.C. is $t(0) = 1$.

29a. To show that $\phi(t) = e^{2t}$ is a solution of the D.E., take its derivative and substitute into the D.E.

30. $[c\, \phi(t)]' + p(t)\, [c\phi(t)] = c[\phi'(t) + p(t)\phi(t)] = 0$ since $\phi(t)$ satisfies the given D.E.

31. $[y_1(t) + y_2(t)]' + p(t)\,[y_1(t) + y_2(t)] =$
    $y_1'(t) + p(t)y_1(t) + y_2'(t) + p(t)y_2(t) = 0 + g(t)$.

32. This problem demonstrates the central idea of the method of variation of parameters for the simplest case. The solution (ii) of the homogeneous D.E. is extended to the corresponding nonhomogeneous D.E. by replacing the constant A by a function A(t), as shown in (iii).

33. Assume $y(t) = A(t)\exp(-\int(-2)\,dt) = A(t)e^{2t}$.
    Differentiating y(t) and substituting into the D.E. yields $A'(t) = t^2$ since the terms involving A(t) add to zero. Thus $A(t) = t^3/3 + c$, which substituted into y(t) yields the solution.

34. $y(t) = A(t)\exp(-\int\dfrac{dt}{t}) = A(t)/t$.

Section 2.2, Page 30

Problems 1 through 8 follow the pattern of solution from Section 2.1

2. Remember to divide both sides of the D.E. by x^2 to get
   $$\mu(t) = \exp(\int\dfrac{3dt}{t}) = e^{3\ln t} = t^3.$$

5. $\mu(t) = \exp(\int \tan t\, dt) = \exp(\int\dfrac{\sin t\, dt}{\cos t})$
   $= \exp[(-\ln(\cos t))] = \exp[\ln(1/\cos t)]$
   $= \sec t$.
   Multiplying both sides of the D.E. by sect and simplifying yields $(y\sec t)' = 2t\sin t$. To solve this the right side must be integrated by parts with $u = t$ and $dv = \sin t\, dt$. Thus $y\sec t = -2t\cos t + 2\sin t + c$, which yields the solution. For all steps we must have $-\pi/2 < t < \pi/2$ for the functions to be defined.

8. $\mu(t) = \exp(-\int\dfrac{2t\,dt}{1+t^2}) = \exp(-\ln(1+t^2)) = \dfrac{1}{1 + t^2}$. Thus
   $(\dfrac{y}{1 + t^2})' = \dfrac{2t}{1 + t^2}$ or $\dfrac{y}{1 + t^2} = \ln(1+t^2) + c$ and
   $y(t) = (1+t^2)\,\ln(1+t^2) + c(1+t^2)$.

In problems 9 through 16, we must determine the largest
interval in which the functions p and g are continuous and
which contain the initial point, and also determine the
constant c of the general solution. The procedure follows
Example 1 of this section.

9a. Writing the D.E. in the form of Eq.(1) of this section we
     have $y' + (2/t)y = t-1 + 1/t$, $y(1) = 1/2$. Thus $p(t)$ and
     $g(t)$ are continuous on any interval not containing the
     origin. Since the initial point is 1, the solution will
     be valid on $0 < t < \infty$. $\mu(t) = t^2$ and thus
     $(t^2 y)' = t^3 - t^2 + t$ and $y = \dfrac{1}{4}t^2 - \dfrac{1}{3}t + \dfrac{1}{2} + \dfrac{c}{t}$.
     Substituting $t = 1$ and $y = 1/2$, we obtain $c = 1/12$, which
     gives the desired solution.

9c. Note that the interval of validity of the actual solution
     agrees with that predicted by Theorem 2.2.1.

9d. As $t \to \infty$ both t^2 and t become unbounded. Note that if
     the initial value for $y(1)$ were such that $c = 0$, then y
     remains finite as $t \to 0$.

11a. $p(t) = \cot t$ and $g(t) = 2\csc t$ are both continuous for
     $n\pi < t < (n+1)\pi$, where n is any integer. Since the
     initial point is $t = \pi/2$, we choose $n = 0$ and conclude
     that the solution will be valid on $0 < t < \pi$. Now
     $\mu(t) = \exp(\int \cot t\, dt = \sin t$ and thus $(y\sin t)' = 2$, which
     gives the general solution $y = (2t + c)/\sin t$. Setting
     $t = \pi/2$ and $y = 1$ we find $c = 1-\pi$.

11c. Since $\sin t$ is zero at $t = 0$ and $t = \pi$, the solution is
     valid for $0 < t < \pi$, as predicted in part a.

14. $p(t) = \dfrac{2(1+t)}{t(2+t)}$ and $g(t) = \dfrac{1+3t}{t(2+t)}$. Thus $p(t)$ and $g(t)$
     are continuous on $-\infty < t < -2$, $-2 < t < 0$ and $0 < t < \infty$.
     Since we are given $y(-1) = 1$, the solution will be valid
     on $-2 < t < 0$. $\mu(t) = t^2 + 2x$.

15. $p(t)$ and $g(t)$ are continuous for all t and thus so is the
     solution. $\mu(t) = e^t$.

16. $\mu(t) = (1 - t^2)^{1/2}$.

$t - 1 + \frac{1}{t}e$

$t^3 - t^2 + t$

17. $\mu(t) = t^2$

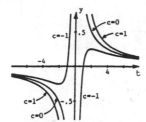

18. $\mu(t) = 1/t$

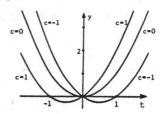

19. $\mu(t) = 1/t$. Solution
       exists only for $t > 0$
       since $g(t) = t^{1/2}$ and
       $p(t) = 1/t$ are defined
       and continuous only for
       $t > 0$.

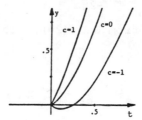

20. $\mu(t) = t$

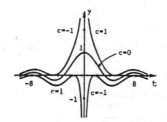

21. If the equation is written in the form of Eq.(1), then
    $p(t) = (\ln t)/(t-3)$ and $g(t) = 2t/(t-3)$. These are defined
    and continuous on the intervals $(0,3)$ and $(3,\infty)$, but
    since the initial point is $t = 1$, the solution will be
    continuous on $0 < t < 3$.

25. $(e^{-t}y)' = e^{-t} + 3e^{-t}\sin t$ so

    $e^{-t}y = -e^{-t} - 3e^{-t}(\dfrac{\sin t + \cos t}{2}) + c$ or

    $y(t) = -1 - (\dfrac{3}{2})e^{-t}(\sin t + \cos t) + ce^{t}$.

    If $y(t)$ is to remain bounded, we must have $c = 0$. Thus

    $y(0) = -1 - \dfrac{3}{2} + c = y_0$ or $c = y_0 + \dfrac{5}{2} = 0$ and $y_0 = -\dfrac{5}{2}$.

27c. The limit as $t \to \infty$ of $[\dfrac{\sqrt{\pi}}{2} \text{erf}(t)+y_0]$ must be zero for

$\lim\limits_{t\to\infty} y(t)$ to be finite. Since $\lim\limits_{t\to\infty} \text{erf}(t) = 1$ we conclude

that $y_0 = -\sqrt{\pi}/2$ and hence

$$\lim_{t\to\infty} e^{t^2}[\frac{\sqrt{\pi}}{2}\ \text{erf}(t)-\frac{\sqrt{\pi}}{2}] = \lim_{t\to\infty} \frac{\frac{\sqrt{\pi}}{2}[\text{erf}(t)-1]}{e^{-t^2}}.$$

L'Hopital's Rule now applies:

$$\lim_{t\to\infty} y(t) = \lim_{t\to\infty} \frac{e^{-t^2}}{-2te^{-t^2}},\ \text{since the derivative of erf}(t)$$

is $\dfrac{2}{\sqrt{\pi}}e^{-t^2}$ from Eq. (20). This last quotient has a limit

of zero.

28c. $y(t) = \sqrt{\dfrac{\pi}{2}}\ e^{t^2/2}\ \text{erf}(t/\sqrt{2}) + t + y_0 e^{t^2/2}.$

30. $\mu(t) = e^{at}$ so the D.E. can be written as
$(e^{at}y)' = be^{at}e^{-\lambda t} = be^{(a-\lambda)t}$. If $a \neq \lambda$, then integration
and solution for y yields $y = [b/(a-\lambda)]e^{-\lambda t} + ce^{-at}$. Then
$\lim\limits_{x\to\infty} y$ is zero since both λ and a are positive numbers.
If $a = \lambda$, then the D.E. becomes $(e^{at}y)' = b$, which yields
$y = (bt+c)/e^{\lambda t}$ as the solution. L'Hopital's Rule gives
$$\lim_{t\to\infty} y = \lim_{t\to\infty} \frac{(bt+c)}{e^{\lambda t}} = \lim_{t\to\infty} \frac{b}{\lambda e^{\lambda t}} = 0.$$

32. There is no unique answer for this situation. One
possible response is to assume $y(t) = ce^{-2t} + 3 - t$, then
$y'(t) = -2ce^{-2t} - 1$ and thus $y' + 2y = 5 - 2t$.

35. $\mu(t) = e^{2t}$. Since g(t) is continuous on the interval
$0 \leq t \leq 1$ we may solve the I.V.P.
$y_1' + 2y_1 = 1$, $y_1(0) = 0$ on that interval to obtain
$y_1 = 1/2 - (1/2)e^{-2t}$, $0 \leq t \leq 1$. g(t) is also continuous
for $1 < t$; and hence we may solve $y_2' + 2y_2 = 0$ to obtain
$y_2 = ce^{-2t}$, $1 < t$. The solution y of the original I.V.P.
must be continuous (since its derivative must exist) and
hence we need c in y_2 so that y_2 at 1 has the same value
as y_1 at 1. Thus

$ce^{-2} = 1/2 - e^{-2}/2$ or $c = (1/2)(e^2-1)$ and we obtain

$$y = \begin{cases} 1/2 - (1/2)e^{-2t} & 0 \le t \le 1 \\ 1/2(e^2-1)e^{-2t} & 1 \le t \end{cases} \qquad \text{and}$$

$$y' = \begin{cases} e^{-2t} & 0 \le t \le 1 \\ (1-e^2)e^{-2t} & 1 < t. \end{cases}$$

Evaluating the two parts of y' at $t_0 = 1$ we see that they are different, and hence y' is not continuous at $t_0 = 1$.

37a. For $n = 0,1$, the D.E. is linear and Eqs.(3) and (4) apply.

37b. Let $v = y^{1-n}$ then $\dfrac{dv}{dt} = (1-n)y^{-n}\dfrac{dy}{dt}$ so $\dfrac{dy}{dt} = \dfrac{1}{1-n}y^n\dfrac{dv}{dt}$, which makes sense when $n \ne 0,1$. Substituting into the D.E. yields $\dfrac{y^n}{1-n}\dfrac{dv}{dt} + p(t)y = q(t)y^n$ or

$v' + (1-n)p(t)y^{1-n} = (1-n)q(t)$. Setting $v = y^{1-n}$ then yields a linear D.E. for v.

38. $n = 3$ so $v = y^{-2}$ and $\dfrac{dv}{dt} = -2y^{-3}\dfrac{dy}{dt}$ or $\dfrac{dy}{dt} = -\dfrac{1}{2}y^3\dfrac{dv}{dt}$.
Substituting this into the D.E. gives
$-\dfrac{1}{2}y^3\dfrac{dv}{dt} + \dfrac{2}{t}y = \dfrac{1}{t^2}y^3$. Simplifying and setting
$y^{-2} = v$ then gives the linear D.E.
$v' - \dfrac{4}{t}v = -\dfrac{2}{t^2}$, where $\mu(t) = \dfrac{1}{t^4}$ and

$v(t) = ct^4 + \dfrac{2}{5t} = \dfrac{2+5ct^5}{5t}$. Thus $y = \pm[5t/(2+5ct^5)]^{1/2}$.

39. $n = 2$ so $v = y^{-1}$ and $\dfrac{dv}{dt} = -y^2\dfrac{dv}{dt}$. Thus the D.E.

becomes $-y^2\dfrac{dv}{dt} - ry = -ky^2$ or $\dfrac{dv}{dt} + rv = k$. Hence

$\mu(t) = e^{rt}$ and $v = k/r + ce^{-rt}$. $y = 1/v$ then yields the solution.

Section 2.3, Page 38

Problems 1 through 20 follow the pattern of the examples
worked in this section. The first eight problems, however,
do not have I.C. so the integration constant, c, cannot be
found.

1.   Write the equation in the form $y\,dy = x^2 dx$. Integrating
     the left side with respect to y and the right side with
     respect to x yields
     $$\frac{y^2}{2} = \frac{x^3}{3} + C, \text{ or } 3y^2 - 2x^3 = c.$$

4.   For $y \neq -3/2$ multiply both sides of the equation by
     $3 + 2y$ to get the separated equation
     $(3+2y)\,dy = (3x^2-1)\,dx$.   Integration then yields
     $3y + y^2 = x^3 - x + c$.

6.   Separating the variables we get $(1-y^2)^{-1/2}\,dy = x^{-1}\,dx$.
     Integrating each side yields $\arcsin y = \ln|x|+c$, so
     $y = \sin[\ln|x|+c]$, $x \neq 0$.

10a. Separating the variables we get $y\,dy = (1-2x)\,dx$, so
     $$\frac{y^2}{2} = x - x^2 + c. \quad \text{Setting } x = 0 \text{ and } y = -2 \text{ we have } 2 = c$$
     and thus $y^2 = 2x - 2x^2 +$ or $y = -\sqrt{2x - 2x^2 + 4}$.   The
     negative square root must be used since $y(0) = -2$.

10c. Rewriting $y(x)$ as $-\sqrt{2(2-x)(x+1)}$, we see that y is
     defined for $-1 \leq x \leq 2$, However, since y' does not exist
     for $x = -1$ or $x = 2$, the solution is valid only for the
     open interval $-1 < x < 2$.

13. Separate variables by factoring the denominator of the
     right side to get $y\,dy = \dfrac{2x}{1+x^2}\,dx$.   Integration yields
     $y^2/2 = \ln(1+x^2)+c$ and use of the I.C. gives $c = 2$.   Thus
     $y = \pm\,[2\ln(1+x^2)+4]^{1/2}$, but we must discard the plus
     square root because of the I.C.   Since $1 + x^2 > 0$, the
     solution is valid for all x.

15. Separating variables and integrating yields
     $y + y^2 = x^2 + c$. Setting $y = 0$ when $x = 2$ yields $c = -4$
     or $y^2 + y = x^2-4$. To solve for y complete the square on

the left side by adding 1/4 to both sides. This yields

$y^2 + y + \dfrac{1}{4} = x^2 - 4 + \dfrac{1}{4}$ or $(y + \dfrac{1}{2})^2 = x^2 - 15/4$. Taking

the square root of both sides yields

$y + \dfrac{1}{2} = \pm\sqrt{x^2 - 15/4}$, where the positive square root

must be taken in order to satisfy the I.C. Thus

$y = -\dfrac{1}{2} + \sqrt{x^2 - 15/4}$, which is defined for $x^2 \geq 15/4$ or

$x \geq \sqrt{15}/2$. The possibility that $x < -\sqrt{15}/2$ is
discarded due to the I.C.

17a. Separating variables gives $(2y-5)dy = (3x^2-e^x)dx$ and
integration then gives $y^2 - 5y = x^3 - e^x + c$. Setting
$x = 0$ and $y = 1$ we have $1 - 5 = 0 - 1 + c$, or $c = -3$.
Thus $y^2 - 5y - (x^3-e^x-3) = 0$ and using the quadratic
formula then gives

$y(x) = \dfrac{5 \pm \sqrt{25+4(x^3-e^x-3)}}{2} = -\dfrac{5}{2} - \sqrt{\dfrac{13}{4} + x^3 - e^x}$. The

negative square root is chosen due to the I.C.

17c. The interval of definition for y must be found
numerically. Approximate values can be found by plotting

$y_1(x) = \dfrac{13}{4} + x^3$ and $y_2(x) = e^x$ and noting the values of x

where the two curves cross.

19a. As above we start with $\cos 3y\,dy = -\sin 2x\,dx$ and integrate

to get $\dfrac{1}{3}\sin 3y = \dfrac{1}{2}\cos 2x + c$. Setting $y = \pi/3$ when

$x = \pi/2$ (from the I.C.) we find that $0 = -\dfrac{1}{2} + c$ or

$c = \dfrac{1}{2}$, so that $\dfrac{1}{3}\sin 3y = \dfrac{1}{2}\cos 2x + \dfrac{1}{2} = \cos^2 x$ (using the

appropriate trigonometric identity). To solve for y we
must choose the branch that passes through the point
$(\pi/2, \pi/3)$ and thus $3y = \pi - \arcsin(3\cos^2 x)$, or

$y = \dfrac{\pi}{3} - \dfrac{1}{3}\arcsin(3\cos^2 x)$.

19c. The solution in part a is defined only for
$0 \leq 3\cos^2 x \leq 1$, or $-\sqrt{1/3} \leq \cos x \leq \sqrt{1/3}$. Taking the
indicated square roots and then finding the inverse
cosine of each side yields $.9553 \leq x \leq 2.1863$, or
$|x-\pi/2| \leq 0.6155$, as the approximate interval.

21. We have $(3y^2-6y)dy = (1+3x^2)dx$ so that $y^3-3y^2 =$
    $x + x^3 - 2$, once the I.C. are used. $dx/dy = 0$ implies
    $3y^2 - 6y = 0$, or $y = 0,2$. For $y = 0$ we have
    $x^3 + x - 2 = 0$, which is satisfied for $x = 1$, which is
    the only zero of the function $w = x^3 + x - 2$. Likewise,
    for $y = 2$, $x = -1$.

23. Separating variables gives $y^{-2}dy = (2+x)dx$, so
    $-y^{-1} = 2x + \dfrac{x^2}{2} + c$. $y(0) = 1$ yields $c = -1$ and thus
    $$y = \frac{-1}{\dfrac{x^2}{2} + 2x - 1} = \frac{2}{2 - 4x - x^2}.$$ This gives
    $\dfrac{dy}{dx} = \dfrac{8 + 4x}{(2-4x-x^2)^2}$, so the minimum value is attained at
    $x = -2$. Note that the solution is defined for
    $-2 - \sqrt{6} < x < -2 + \sqrt{6}$ and has vertical asymptotes at
    the end points of the interval.

25. Separating variables and integrating yields $3y + y^2 =$
    $\sin 2x + c$. $y(0) = -1$ gives $c = -2$ so that
    $y^2 + 3y + (2-\sin 2x) = 0$. The quadratic formula then
    gives $y = -\dfrac{3}{2} + \sqrt{\sin 2x + 1/4}$, which is defined for
    $-.126 < x < 1.697$ (found by solving $\sin 2x = -.25$ for x
    and noting $x = 0$ is the initial point). Thus we have
    $\dfrac{dy}{dx} = \dfrac{\cos 2x}{(\sin 2x + \dfrac{1}{4})}$, which yields $x = \pi/4$ as the only
    critical point in the above interval. Using the second
    derivative test or graphing the solution clearly
    indicates the critical point is a maximum.

27a. By sketching the direction field and using the D.E. we
    note that $y' < 0$ for $y > 4$ and y' approaches zero as y
    approaches 4. For $0 < y < 4$, $y' > 0$ and again approaches
    zero as y approaches 4. Thus $\lim\limits_{t\to\infty} y = 4$ if $y_0 > 0$. For
    $y_0 < 0$, $y' < 0$ for all y and hence y becomes negatively
    unbounded $(-\infty)$ as t increases. If $y_0 = 0$, then $y' = 0$
    for all t, so $y = 0$ for all t.

27b. Separating variables and using a partial fraction

expansion we have $(\frac{1}{y} - \frac{1}{y-4})dy = \frac{4}{3}tdt$. Hence

$\ln|\frac{y}{y-4}| = \frac{2}{3}t^2 + c_1$ and thus $|\frac{y}{y-4}| = e^{c_1}e^{2t^2/3} = ce^{2t^2/3}$,

where c is positive. For $y_0 = .5$ this becomes

$\frac{y}{4-y} = ce^{2t^2/3}$ and thus $c = \frac{.5}{3.5} = \frac{1}{7}$. Using this value

for c and solving for y yields $y(t) = \dfrac{4}{1 + 7e^{-2t^2/3}}$.

Setting this equal to 3.98 and solving for t yields
t = 3.2953.

29. Rewrite the D.E. as $\dfrac{dx}{dy} = \dfrac{cy + d}{ay + b}$ and assume $a \neq 0$ and

$ay + b \neq 0$. Then $dx = (\dfrac{c}{a} + \dfrac{ad-bc}{a(ay+b)})dy$. Integration then

yields the desired answer.

30. If $v = y/x$ then $y = vx$ and $\dfrac{dy}{dx} = v + x\dfrac{dv}{dx}$ and thus the

D.E. becomes $v + x\dfrac{dv}{dx} = \dfrac{vx-4x}{x-vx} = \dfrac{v-4}{1-v}$. Subtracting v

from both sides yields $x\dfrac{dv}{dx} = \dfrac{v^2-4}{1-v}$, which separates into

$\dfrac{1-v}{v^2-4}dv = \dfrac{1}{x}dx$. To integrate the left side use partial

fractions to write $\dfrac{1-v}{v-4} = \dfrac{A}{v-2} + \dfrac{B}{v+2}$, which yields

A = -1/4 and B = -3/4. Integration then gives

$-\dfrac{1}{4}\ln|v-2| - \dfrac{3}{4}\ln|v+2| = \ln|x| - k$, or

$\ln|x^4||v-2||v+2|^3 = 4k$ after manipulations using
properties of the ln function. Setting $v = y/x$ and
further algebraic manipulations yield

$|y-2x||y+2x|^3 = c$, where $c = e^{4k}$.

31a. The maximum and minimum values of $y/(1+2y^2)$ occur at
$y = +\sqrt{1/2}$ and $y = -\sqrt{1/2}$ respectively. Thus
$-\dfrac{1}{2\sqrt{2}} \leq \dfrac{y}{1+2y^2} \leq \dfrac{1}{2\sqrt{2}}$ and we know that $|\cos x| \leq 1$, hence
f(x,y) is bounded as stated.

31b. Since $dy/dx = f(x,y)$, we have from part (a)

$-\dfrac{1}{2\sqrt{2}} \leq \dfrac{dy}{dx} \leq \dfrac{1}{2\sqrt{2}}$. Separating variables and using

the format of Eq.(15) of this section yields

$-\int_1^\phi \dfrac{dt}{2\sqrt{2}} \leq \int_1^\phi dy \int_0^x \dfrac{dt}{2\sqrt{2}}$. Integration and evaluation

gives $\dfrac{-x}{2\sqrt{2}} \leq \phi(x) - 1 \leq \dfrac{x}{2\sqrt{2}}$ or $|\phi(x) - 1| \leq \dfrac{|x|}{2\sqrt{2}}$ for

all x. Geometrically, this says that the solution is

bounded by two straight lines of slope $\pm\dfrac{1}{2\sqrt{2}}$ passing

through the initial point $x = 0$, $y = 1$. Thus the
solution is bounded for all finite values of x.

Section 2.4, Page 45

2.   Theorem 2.4.1 guarantees a unique solution to the D.E.
through any point (t_0, y_0) such that $t_0^2 + y_0^2 < 1$ since
$\dfrac{\partial f}{\partial y} = -y(1-t^2-y^2)^{1/2}$ is defined and continuous only for
$1-t^2-y^2 > 0$. Note also that $f = (1-t^2-y^2)^{1/2}$ is defined
and continuous in this region as well as on the boundary
$t^2+y^2 = 1$. The boundary can't be included in the final
region due to the discontinuity of $\dfrac{\partial f}{\partial y}$ there.

3.   We are guaranteed a solution to the D.E. passing through
any point in the ty plane since $f = 2ty/(1+y^2)$ and $\dfrac{\partial f}{\partial y} = $
$2t(1-y^2)/(1+y^2)^2$ are defined and continuous for all t and
y.

7.   In this case $f = \dfrac{1+t^2}{y(3-y)}$ and $\dfrac{\partial f}{\partial y} = \dfrac{1+t^2}{y(3-y)^2} - \dfrac{1+t^2}{y^2(3-y)}$,
which are both continuous everywhere except for $y = 0$ and
$y = 3$.

9.   The D.E. may be written as $ydy = -4tdt$ so that
$\dfrac{y^2}{2} = -2t^2+c$, or $y^2 = c-4t^2$. The I.C. then yields
$y_0^2 = c$, so that $y^2 = y_0^2 - 4t^2$ or $y = \pm\sqrt{y_0^2-4t^2}$, which is

defined for $4t^2 < y_0^2$ or $|t| < |y_0|/2$. Note that $y_0 \neq 0$
since Theorem 2.4.1 does not hold there.

13. From the direction field and the given D.E. it is noted
that for $t > 0$ and $y < 0$ that $y' < 0$, so $y \to -\infty$ for
$y_0 < 0$. Likewise, for $0 < y_0 < 3$, $y' > 0$ and $y' \to 0$ as
$y \to 3$, so $y \to 3$ for $0 < y_0 < 3$ and for $y_0 > 3$, $y' < 0$
and again $y' \to 0$ as $y \to 3$, so $y \to 3$ for $y_0 > 3$. For
$y_0 = 3$, $y' = 0$ and $y = 3$ for all t.

18a. For $y_1 = 1-t$, $y_1' = -1 = \dfrac{-t+[t^2+4(1-t)]^{1/2}}{2}$

$$= \frac{-t+[(t-2)^2]^{1/2}}{2}$$

$$= \frac{-t+|t-2|}{2} = -1 \text{ if}$$

$(t-2) \geq 0$, by the definition of absolute value. Setting
$t = 2$ in y_1 we get $y_1(2) = -1$, as required.

18b. By Theorem 2.4.1 we are guaranteed a unique solution only
where $f(t,y) = \dfrac{-t+(t^2+4y)^{1/2}}{2}$ and $f_y(t,y) = (t^2+4y)^{-1/2}$ are
continuous. In this case the initial point $(2,-1)$ lies
in the region $t^2 + 4y \leq 0$, in which case $\dfrac{\partial f}{\partial y}$ is not
continuous and hence the theorem is not applicable and
there is no contradiction.

18c. If $y = y_2(t)$ then we must have $ct + c^2 = -t^2/4$, which is
not possible since c must be a constant.

Section 2.5, Page 54

1a. Comparing this problem with Eq.(1), we see that
$r = .0525$ and thus Eq.(8) yields $\tau = \ln 2/.0525 = 13.20$
years as the half life of plutonium 241.

1b. Solving $dQ/dt = -0.0525Q$ with $Q(0) = 50$ mg we find
$Q(t) = 50e^{-0.0525t}$. Thus $Q(10) = 50e^{-0.525} = 29.6$ mg.

4a. From Ex.(1) $r = 0.02828$ and thus $dQ/dt = -0.01818Q + 1$
gives the rate of change of thorium 234 in mg/day. The
solution of this equation with the initial condition of

Q(0) = 100 mg can be found using methods of Section 2.1
and is $Q(t) = (100 - \dfrac{1}{0.02828})e^{-0.02828t} + \dfrac{1}{0.02828}$, or

$Q(t) = 64.64e^{-0.02828t} + 35.36$ mg.

4c. Setting Q(t) = 35.86 mg in the answer to part(a) and
solving for t yields $t = \dfrac{-\ln(.5/64.64)}{0.02828} = 171.9$ days.

8a. Set $S_0 = 0$ in Eq.(16) (or solve Eq.(15) with S(0) = 0).

8b. Set r = .075, t = 40 and S(t) = \$1,000,000 in the answer
to (a) and then solve for k.

8c. Set k = \$2,000, t = 40 and S(t) = \$1,000,000 in the
answer to (a) and then solve numerically for r.

10. The rate of accumulation due to interest is .1S and the
rate of decrease is k dollars per year and thus
dS/dt = .1S - k, S(0) = \$8,000. Solving this for S(t)
yields $S(t) = 8000e^{.1t} - 10k(e^{.1t}-1)$ and substitution of
t = 3 gives k = \$3,086.64 per year.

11. Since we are assuming continuity, either convert the
monthly payment into an annual payment or convert the
yearly interest rate into a monthly interest rate for 240
months. Then proceed as in Prob. 10.

13a. Using Eq. (15) we have $\dfrac{dS}{dt} = \dfrac{.09}{12}S - 800(1+\dfrac{t}{120})$ or

$\dfrac{dS}{dt} - \dfrac{3}{400}S = -(800+\dfrac{20}{3}t)$, S(0) = 100,000. Using an

integrating factor and integration by parts (or using a

D.E. solver) we get $S(t) = \dfrac{6,080,000}{27} + \dfrac{800}{9}t + ce^{3t/400}$.

Using the I.C. yields $c = \dfrac{-3,380,000}{27}$. Substituting this

value into S, setting S(t) = 0, and solving numerically
for t yields t ≅ 135.363 months.

15. Let p(t) be the population at any time and let k be the
constant of proportionality. Then the rate of change of
p(t) is given by dp/dt = kp, with $p(0) = 6 \times 10^8$ and
$p(300) = 2.8 \times 10^9$. The solution of the D.E. is
$p(t) = ce^{kt}$, where c is found by setting t = 0:

$p(0) = c = 6 \times 10^8$. Thus $p(t) = (6 \times 10^8)e^{kt}$. To find k, set $t = 300$ and $p(300) = 2.8 \times 10^9$, yielding: $e^{300k} = (2.8/6)10$ or $k = .005135$. Hence $p(t) = (6 \times 10^8)e^{.005135t}$ gives the population at any time t. Setting $p(t) = 2.5 \times 10^{10}$ and solving for t yields the number of years after 1650 when the greatest population is reached.

17a. This problem can be done numerically using appropriate D.E. solver software. Analytically we have $\dfrac{dy}{y} = (.1 + .2\sin t)dt$ by separating variables and thus $y(t) = c\exp(.1t - .2\cos t)$. $y(0) = 1$ gives $c = e^{.2}$, so $y(t) = \exp(.2 + .1t - .2\cos t)$. Setting $y = 2$ yields $\ln 2 = .2 + .1\tau - .2\cos\tau$, which can be solved numerically to give $\tau = 2.9632$. If $y(0) = y_0$, then as above, $y(t) = y_0 \exp(.2 + .1t - .2\cos t)$. Thus if we set $y = 2y_0$ we get the same numerical equation for τ and hence the doubling time has not changed.

19. From Eq.(26) we have $190° = 70° + (200° - 70°)e^{-k}$ or $k = -\ln\dfrac{120}{130} = \ln(13/12)$. Hence if $\theta(T) = 150°$ we get: $150° = 70° + 130°e^{-\ln(13/12)T}$. Solving for T yields $T = \ln(8/13)/-\ln(13/12) = \ln(13/8)/\ln(13/12)$.

21. To determine the cooling rate, the information relevant to the morgue is used in Eq.(26): $\theta(t) = 40° + (85° - 40°)e^{-kt}$. Setting $t = 1$ and $\theta(1) = 60$ and solving for k yields $k = .8109$ (hour)$^{-1}$. To find the time of death, use Eq.(22) with $T = 70°$ and $\theta_0 = 98.6°$: $\theta(t) = 70° + 28.6e^{-.8109t}$. Setting $t = T$ and $\theta(T) = 85$ and solving for T gives $T = .7959$ hours before midnight, or 11:12 P.M.

22. The D.E. expressing the evaporation is $dV/dt = kS$, where the volume $V = \dfrac{4}{3}\pi r^3$ and the surface area $S = 4\pi r^2$. The D.E., in terms of r, then is $r^2\dfrac{dr}{dt} = kr^2$, with $r(0) = 3$ and $r(1) = 2$. When $r \neq 0$ the D.E. becomes $r' = k$, or $r(t) = kt + c$. Recall that r must be positive for the problem to have physical meaning.

23. Let $S(t)$ be the amount of salt that is present at any
    time t, then $S(0) = 0$ is the original amount of salt in
    the tank, 2γ is the amount of salt entering per minute,
    and $2(S/120)$ is the amount of salt leaving per minute
    (all amounts measured in grams). Thus
    $dS/dt = 2\gamma - 2S/120$, $S(0) = 0$.

25. We must first find the amount of salt that is present
    after 10 minutes. For the first 10 minutes (if we let
    $Q(t)$ be the amount of salt in the tank):
    $$\frac{dQ}{dt} = \frac{1}{2}(2) - 2\frac{Q(t)}{100}, \quad Q(0) = 0. \quad \text{This I.V.P. has the}$$
    solution: $Q(10) = 50(1-e^{-.2}) = 9.063$ lbs. of salt in the
    tank after the first 10 minutes. At this point no more
    salt is allowed to enter, so the new I.V.P. (letting $P(t)$
    be the amount of salt in the tank) is:
    $$\frac{dP}{dt} = (0)(2) - 2\frac{P(t)}{100}, \quad P(0) = Q(10) = 9.063. \quad \text{The}$$
    solution of this problem is $P(t) = 9.063e^{-.02t}$, which
    yields $P(10) = 7.42$ lbs.

26. Salt flows into the tank at the rate of $(1)(3)$ lb/min.
    and it flows out of the tank at the rate of $\dfrac{Q(t)}{200+t}(2)$
    lb/min. since the volume of water in the tank at any time
    t is $200 + (1)(t)$ gallons (due to the fact that water
    flows into the tank faster than it flows out). Thus the
    I.V.P. is $dQ/dt = 3 - \dfrac{2}{200+t}Q(t)$, $Q(0) = 100$.

27. Hint: let $Q(t)$ be the quantity of carbon monoxide in the
    room at any time t. Then the concentration is given by
    $x(t) = Q(t)/1200$.

28a. The required I.V.P. is $dQ/dt = kr + P - \dfrac{Q(t)}{V}r$,
    $Q(0) = Vc_0$. Since $c = Q(t)/V$, the I.V.P. may be
    rewritten $Vc'(t) = kr + P - rc$, $c(0) = c_0$, which has the
    solution $c(t) = k + \dfrac{P}{r} + (c_0 - k - \dfrac{P}{r})e^{-rt/V}$.

28b. Set $k = 0$, $P = 0$, $t = T$ and $c(T) = .5c_0$ in the solution
    found in (a).

Problems 1 through 13 follow the pattern illustrated in
Fig.2.6.3 and the discussion following Eq.(11).

3. The critical points are found by setting $\frac{dy}{dt}$ equal to
    zero. Thus $y = 0,1,2$ are the critical points. The graph
    of $y(y-1)(y-2)$ is positive for $0 < y < 1$ and $2 < y$ and
    negative for $1 < y < 2$. Thus $y(t)$ is increasing
    $(\frac{dy}{dt} > 0)$ for $0 < y < 1$ and $2 < y$ and decreasing
    $(\frac{dy}{dt} < 0)$ for $1 < y < 2$. Therefore 0 and 2 are unstable
    critical points while 1 is an asymptotically stable
    critical point.

6. $\frac{dy}{dt}$ is zero only when Arctany is zero. $\frac{dy}{dt} > 0$ for
    $y < 0$ and $\frac{dy}{dt} < 0$ for $y > 0$. Thus $y = 0$ is an
    asymptotically stable critical point.

7c. Separate variables to get $\frac{dy}{(1-y)^2} = kt$. Integration
    yields $\frac{1}{1-y} = kt + c$, or $y = 1 - \frac{1}{kt + c} = \frac{kt + c - 1}{kt + c}$.
    Setting $t = 0$ and $y(0) = y_0$ yields $y_0 = \frac{c-1}{c}$ or
    $c = \frac{1}{1-y_0}$. Hence $y(t) = \frac{(1-y_0)kt + y_0}{(1-y_0)kt + 1}$. For $y_0 > 1$
    notice that the denominator will have a zero for some
    value of t, depending on the values chosen for y_0 and k.
    Thus the solution has a discontinuity at that point.

9. Setting $\frac{dy}{dt} = 0$ we find $y = 0, \pm 1$ are the critical
    points. Since $\frac{dy}{dt} > 0$ for $y < -1$ and $y > 1$ while $\frac{dy}{dt} < 0$
    for $-1 < y < 1$ we may conclude that $y = -1$ is
    asymptotically stable, $y = 0$ is semistable, and $y = 1$ is
    unstable.

11. $y = b^2/a^2$ and $y = 0$ are the only critical points. For

$0 < y < b^2/a^2$, $\dfrac{dy}{dt} < 0$ and thus $y = 0$ is asymptotically stable. For $y > b^2/a^2$, $dy/dt > 0$ and thus $y = b^2/a^2$ is unstable.

14. If $f'(y_1) < 0$ then the slope of f is negative at y_1 and thus $f(y) > 0$ for $y < y_1$ and $f(y) < 0$ for $y > y_1$ since $f(y_1) = 0$. Hence y_1 is an asysmtotically stable critical point. A similar argument will yield the result for $f'(y_1) > 0$.

16b. The graph of $\dfrac{dy}{dt}$ vs y has a maximum point at $y = K/e$.

Thus $\dfrac{dy}{dt}$ is positive and increasing for $0 < y < K/e$ and thus $y(t)$ is concave up for that interval. Similarly $\dfrac{dy}{dt}$ is positive and decreasing for $K/e < y < K$ and thus $y(t)$ is concave down for that interval.

16c. $\ln(K/y)$ is very large for small values of y and thus $(ry)\ln(K/y) > ry(1 - y/K)$ for small y. Since $\ln(K/y)$ and $(1 - y/K)$ are both strictly decreasing functions of y and since $\ln(K/y) = (1 - y/K)$ only for $y = K$, we may conclude that $\dfrac{dy}{dt} = (ry)\ln(K/y)$ is never less than $\dfrac{dy}{dt} = ry(1 - y/K)$.

17a. If $u = \ln(y/K)$ then $y = Ke^u$ and $\dfrac{dy}{dt} = Ke^u\dfrac{du}{dt}$ so that the D.E. becomes $du/dt = -ru$.

18b. Use the results of Problem 14.

18d. Differentiate Y with respect to E.

19a. Set $\dfrac{dy}{dt} = 0$ and solve for y using the quadratic formula.

19b. Use the results of Problem 14.

19d. If $h > rK/4$ there are no critical points (see part a) and $\dfrac{dy}{dt} < 0$ for all t.

22a. If $z = x/n$ then $dz/dt = \dfrac{1}{n}\dfrac{dx}{dt} - \dfrac{x}{n^2}\dfrac{dn}{dt}$. Use of

   Equations (i) and (ii) then gives the I.V.P. (iii).

22b. Separate variables to get $\dfrac{dz}{z(1-z)} = -\beta dt$. Using partial

   fractions this becomes $\dfrac{dz}{z} + \dfrac{dz}{1-z} = -\beta dt$. Integration

   and solving for z yields the answer.

24a. Plot dx/dt vs x and observe that $x = p$ and $x = q$ are
   critical points. Also note that $dx/dt > 0$ for
   $x < \min(p,q)$ and $x > \max(p,q)$ while $dx/dt < 0$ for x
   between $\min(p,q)$ and $\max(p,q)$. Thus $x = \min(p,q)$ is an
   asymptotically stable point while $x = \max(p,q)$ is
   unstable. To solve the D.E., separate variables and use

   partial fractions to obtain $\dfrac{1}{q-p}[\dfrac{dx}{q-x} - \dfrac{dx}{p-x}] = \alpha dt$.

   Integration and solving for x yields the solution.

24b. $x = p$ is a semistable critical point and since $\dfrac{dx}{dt} > 0$,

   $x(t)$ is an increasing function. Thus for $x(0) = 0$, $x(t)$
   approaches p as $t \to \infty$. To solve the D.E., separate
   variables and integrate.

Section 2.7, Page 79

1a. If $x(t)$ is the height above the ground, then the I.V.P.
   for $v(t)$ is $dv/dt = -g$, $v(0) = 20$. Thus

   $\dfrac{dx}{dt} = v(t) = 20 - gt$ and $x(t) = 20t - (g/2)t^2 + c$. Since

   $x(0) = 30$, $c = 30$ and $x(t) = 20t - (g/2)t^2 + 30$. At the
   maximum height $v(t_m) = 0$ and thus
   $t_m = 20/9.8 = 2.04$ sec., which when substituted in the
   equation for $x(t)$ yields the maximum height.

1b. At the ground $x(t_g) = 0$ and thus $20t_g - 4.9t_g^2 + 30 = 0$.

2. The I.V.P. in this case is $m\dfrac{dv}{dt} = -\dfrac{1}{30}v - mg$, $v(0) = 20$,

   where the positive direction is measured upward.

5.   The I.V.P. is $m(dv/dt) = mg - kv$, $v(0) = 0$ which has the solution $v(t) = \frac{mg}{k}(1 - e^{-kt/m})$. Setting

$v(t) = .9\frac{mg}{k}$ and solving for t yields the answer.

7a.   The I.V.P. is $m\frac{dv}{dt} = mg - .75v$, $v(0) = 0$ and v is measured positively downward. The solution to this equation is $v(t) = 240(1-e^{-.133t})$ so that $v(10) = 176.7$ ft/sec.

7b.   Integration of $v(t)$ as found in (a) yields $x(t) = 240t + 1800(e^{-.133t}-1)$ where x is measured positively down from the altitude of 5000 feet. Set $t = 10$ to find the distance traveled when the parachute opens.

7c.   After the parachute opens the I.V.P. is $m\frac{dv}{dt} = mg-12v$, $v(0) = 176.7$, which has the solution $v(t) = 161.7e^{-2.133t} + 15$ and where $t = 0$ now represents the time the parachute opens. Letting $t\rightarrow\infty$ yields the limiting velocity.

7d.   Integrate $v(t)$ as found in (c) to find $x(t) = 15t - 75.8e^{-2.133t} + C_2$. $C_2 = 75/8$ since $x(0) = 0$, x now being measured from the point where the parachute opens. Setting $x = 3925.5$ will then yield the length of time the skydiver is in the air after the parachute opens.

8.   The I.V.P. is $m(dv/dt) = mg - k\sqrt{v}$; $v(0) = v_0$ where the positive direction is down. To solve, separate variables to obtain $\frac{mdv}{mg - k\sqrt{v}} = dt$ and let $u = mg - k\sqrt{v}$ to integrate the left side.

9.   The I.V.P. is $m(dv/dt) = mg - kv^2$, $v(0) = 0$. To solve, separate variables and use partial fractions.

12b. From part (a) $v(t) = -\frac{mg}{k} + [v_0 + \frac{mg}{k}]e^{-kt/m}$. As $k \rightarrow 0$ this has the indeterminant form of $-\infty + \infty$. Thus rewrite $v(t)$ as $v(t) = [-mg + (v_0k + mg)e^{-kt/m}]/k$ which has the

indeterminant form of 0/0, as $k \rightarrow 0$ and hence
L'Hopital's Rule may be applied with k as the variable.

13a. The equation of motion is $m(dv/dt) = w-R-B$ which, in this
problem, is

$$\frac{4}{3}\pi a^3 \rho (dv/dt) = \frac{4}{3}\pi a^3 \rho g - 6\pi\mu a v - \frac{4}{3}\pi a^3 \rho' g.$$ The limiting

velocity occurs when $dv/dt = 0$.

13b. Since the droplet is motionless, $v = dv/dt = 0$, we have

the equation of motion $0 = (\frac{4}{3})\pi a^3 \rho g - Ee - (\frac{4}{3})\pi a^3 \rho' g$,

where ρ is the density of the oil and ρ' is the density
of air. Solving for e yields the answer.

14. All three parts can be answered from one solution if k
represents the resistance and if the method of solution
of Example 2 is used. Thus we have

$$m\frac{dv}{dt} = mv\frac{dv}{dx} = mg - kv, \quad v(0) = 0,$$ where we have assumed

the velocity is a function of x. The solution of this
I.V.P. involves a logarithmic term, and thus the answers
to parts (a) and (c) must be found using a numerical
procedure.

15b. Note that 32 $ft/sec^2 = 78,545$ m/hr^2.

16. This problem is the same as Example 2 through Eq.(15).
In this case the I.C. is $v(\xi R) = v_o$, so $c = v_o^2 - \frac{2gR}{1+\xi}$.

v_e is found by noting that $v_o^2 \geq \frac{2gR}{1+\xi}$ in order for v^2 to

always be positive. From Example 2, the escape velocity
for a surface launch is $v_e(0) = \sqrt{2gR}$. We want the
escape velocity of $x_o = R$ to have the relation

$v_e(\xi R) = .85v_e(0)$, which yields $\xi = (0.85)^{-2} - 1 \cong 0.384$.
If R = 4000 miles then $x_o = \xi R = 1536$ miles.

17b. From part a) $\frac{dx}{dt} = v = u\cos A$ and hence

$x(t) = (u\cos A)t + d_1$. Since $x(0) = 0$, we have $d_1 = 0$ and

$x(t) = (u\cos A)t$. Likewise $\frac{dy}{dt} = -gt + u\sin A$ and

therefore $y(t) = -gt^2/2 + (u\sin A)t + d_2$. Since $y(0) = h$

we have $d_2 = h$ and $y(t) = -gt^2/2 + (u\sin A)t + h$.

17d. Let t_w be the time the ball reaches the wall. Then

$x(t_w) = L = (ucosA)t_w$ and thus $t_w = \dfrac{L}{ucosA}$. For the ball

to clear the wall $y(t_w) \geq H$ and thus (setting

$t_w = \dfrac{L}{ucosA}$, $g = 32$ and $h = 3$ in y) we get

$\dfrac{-16L^2}{u^2cos^2A} + LtanA + 3 \geq H.$

17e. Setting $L = 350$ and $H = 10$ we get $\dfrac{-161.98}{cos^2A} + 350\dfrac{sinA}{cosA} \geq 7$

or $7cos^2A - 350cosAsinA + 161.98 \leq 0$. This can be solved
numerically or by plotting the left side as a function of
A and finding where the zero crossings are.

17f. Setting $L = 350$, and $H = 10$ in the answer to part d

yields $\dfrac{-16(350)^2}{u^2cos^2A} + 350tanA = 7$, where we have chosen the

equality sign since we want to just clear the wall.

Solving for u^2 we get $u^2 = \dfrac{1,960,000}{175sin2A-7cos^2A}$. Now u will

have a minimum when the denominator has a maximum. Thus
$350cos2A + 7sin2A = 0$, or $tan2A = -50$, which yields
$A = .7954$ rad. and $u = 106.89$ ft./sec.

Section 2.8, Page 88

3. $M(x,y) = 3x^2-2xy+2$ and $N(x,y) = 6y^2-x^2+3$, so $M_y = -2x = N_x$
     and thus the D.E. is exact. Integrating $M(x,y)$ with
     respect to x we get $\psi(x,y) = x^3 - x^2 + 2x + H(y)$. Taking
     the partial derivative of this with respect to y and
     setting it equal to $N(x,y)$ yields $-x^2+h'(y) = 6y^2-x^2+3$, so
     that $h'(y) = 6y^2 + 3$ and $h(y) = 2y^3 + 3y$. Substitute this
     $h(y)$ into $\psi(x,y)$ and recall that the equation which
     defines $y(x)$ implicitly is $\psi(x,y) = c$. Thus
     $x^3 - x^2y + 2x + 2y^3 + 3y = c$ is the equation that yields
     the solution.

5.   Writing the equation in the form $M(x,y)dx + N(x,y)dy = 0$
     gives $M(x,y) = ax + by$ and $N(x,y) = bx + cy$. Now
     $M_y = b = N_x$ and the equation is exact. Integrating $M(x,y)$
     with respect to x yields $\psi(x,y) = (a/2)x^2 + bxy + h(y)$.

Differentiating ψ with respect to y (x constant) and setting $\psi_y(x,y) = N(x,y)$ we find that h'(y) = cy and thus h(y) = $(c/2)y^2$. Hence the solution is given by

$(a/2)x^2 + bxy + (c/2)y^2 = k$.

7. $M_y(x,y) = e^x\cos y - 2\sin x = N_x(x,y)$ and thus the D.E. is exact. Integrating M(x,y) with respect to x gives
    $\psi(x,y) = e^x\sin y + 2y\cos x + h(y)$. Finding $\psi_y(x,y)$ from this and setting that equal to N(x,y) yields h'(y) = 0 and thus h(y) is a constant. Hence an implicit solution of the D.E. is $e^x\sin y + 2y\cos x = c$. The solution y = 0 is also valid since it satisfies the D.E. for all x.

9. If you try to find $\psi(x,y)$ by integrating M(x,y) with respect to x you must integrate by parts. Instead find $\psi(x,y)$ by integrating N(x,y) with respect to y to obtain $\psi(x,y) = e^{xy}\cos 2x - 3y + g(x)$. Now find g(x) by differentiating $\psi(x,y)$ with respect to x and set that equal to M(x,y), which yields g'(x) = 2x or $g(x) = x^2$.

12. As long as $x^2 + y^2 \neq 0$, we can simplify the equation by multiplying both sides by $(x^2 + y^2)^{3/2}$. This gives the exact equation xdx + ydy = 0. The solution to this equation is given implicitly by $x^2 + y^2 = c$. If you apply Theorem 2.8.1 and its construction without the simplification, you get $(x^2 + y^2)^{-1/2} = C$ which can be written as $x^2 + y^2 = c$ under the same assumption required for the simplification.

14. $M_y = 1$ and $N_x = 1$, so the D.E. is exact. Integrating M(x,y) with respect to x yields
    $\psi(x,y) = 3x^3 + xy - x + h(y)$. Differentiating this with respect to y and setting $\psi_y(x,y) = N(x,y)$ yields
    h'(y) = -4y or $h(y) = -2y^2$. Thus the implicit solution is $3x^3 + xy - x - 2y^2 = c$. Setting x = 1 and y = 0 gives c = 2 so that $2y^2 - xy + (2+x-3x^3) = 0$ is the implicit solution satisfying the given I.C. Use the quadratic formula to find y(x), where the negative square root is used in order to satisfy the I.C.

15. We want $M_y(x,y) = 2xy + bx^2$ to be equal to $N_x(x,y) = 3x^2 + 2xy$. Thus we must have b = 3. This

gives $\psi(x,y) = \frac{1}{2}x^2y^2 + x^3y + h(y)$ and consequently

$h'(y) = 0$. After multiplying through by 2, the solution is given implicitly by $x^2y^2 + 2x^3y = c$.

19. $M_y(x,y) = 3x^2y^2$ and $N_x(x,y) = 1 + y^2$ so the equation is not exact by Theorem 2.8.1. Multiplying by the integrating factor $\mu(x,y) = 1/xy^3$ we get
$x + \frac{(1+y^2)}{y^3}y' = 0$, which is an exact equation since
$M_y = N_x = 0$ (it is also separable). In this case
$\psi = \frac{1}{2}x^2 + h(y)$ and $h'(y) = y^{-3} + y^{-1}$ so that
$x^2 - y^{-2} + 2\ln|y| = c$ gives the solution implicitly.

22. Multiplication of the given D.E. (which is not exact) by
$\mu(x,y) = xe^x$ yields $(x^2 + 2x)e^x\sin y\, dx + x^2e^x\cos y\, dy$,
which is exact since $M_y(x,y) = N_x(x,y) = (x^2+2x)e^x\cos y$.
To solve this exact equation it's easiest to integrate
$N(x,y) = x^2e^x\cos y$ with respect to y to get
$\psi(x,y) = x^2e^x\sin y + g(x)$. Solving for g(x) yields the implicit solution.

23. This problem is similar to the derivation leading up to Eq.(26). Assuming that μ depends only on y, we find from Eq.(25) that $\mu' = Q\mu$, where $Q = (N_x - M_y)/M$ must depend on y alone. Solving this last D.E. yields $\mu(y)$ as given. This method provides an alternative approach to Problems 27 through 30.

25. The equation is not exact so we must attempt to find an integrating factor. Since $\frac{1}{N}(M_y-N_x) = \frac{3x^2 + 2x + 3y^2 - 2x}{x^2 + y^2} = 3$ is a function of x alone there is an integrating factor depending only on x, as shown in Eq.(26). Then $d\mu/dx = 3\mu$, and the integrating factor is $\mu(x) = e^{3x}$. Hence the equation can be solved as in Example 4.

26. An integrating factor can be found which is a function of x only, yielding $\mu(x) = e^{-x}$. Alternatively, you might recognize that $y' - y = e^{2x} - 1$ is a linear first order equation which can be solved as in Section 2.1.

27. Using the results of Problem 23, it can be shown that
$\mu(y) = y$ is an integrating factor. Thus multiplying the
D.E. by y gives $y\,dx + (x - y\sin y)\,dy = 0$, which can be
identified as an exact equation. Alternatively, one can
rewrite the last equation as $(y\,dx + x\,dy) - y\sin y\,dy = 0$.
The first term is $d(xy)$ and the last can be integrated by
parts. Thus we have $xy + y\cos y - \sin y = c$.

29. By multiplying by $\sin y$ we obtain
$e^x\sin y\,dx + e^x\cos y\,dy + 2y\,dy = 0$, and the first two
terms are just $d(e^x\sin y)$. Thus $e^x\sin y + y^2 = c$.

31. Using the results of Problem 24, it can be shown that
$\mu(xy) = xy$ is an integrating factor. Thus multiplying by
xy we have $(3x^2y + 6x)\,dx + (x^3 + 3y^2)\,dy = 0$, which can be
identified as an exact equation. Alternatively, we can
observe that the above equation can be written as
$d(x^3y) + d(3x^2) + d(y^3) = 0$, so that $x^3y + 3x^2 + y^3 = c$.

Section 2.9, Page 93

Problems 1 through 14 are done in the same manner as the
illustrative example of this section.

1. The D.E. can be written as $\dfrac{dy}{dx} = 1 + y/x$ and is thus

    homogeneous. Setting $v = y/x$ yields $x\dfrac{dv}{dx} + v = 1 + v$ so

    that $\dfrac{dv}{dx} = \dfrac{1}{x}$. This equation is separable and has the

    solution $v = \ln|x| + c$. Since $v = y/x$, we obtain
    $y = x\ln|x| + cx$ as the solution.

3. Simplifying the right side of the D.E. gives
    $dy/dx = 1 + (y/x) + (y/x)^2$ so the equation is
    homogeneous. The substitution $y = vx$ leads to

    $v + x\dfrac{dv}{dx} = 1 + v + v^2$ or $\dfrac{dv}{1 + v^2} = \dfrac{dx}{x}$. Solving, we get

    $\arctan v = \ln|x| + c$. Substituting for v we obtain
    $\arctan(y/x) - \ln|x| = c$.

5. Dividing the numerator and denominator of the right side
    by x and substituting $y = vx$ we get $v + x\dfrac{dv}{dx} = \dfrac{4v - 3}{2-v}$

which can be rewritten as $x\dfrac{dv}{dx} = \dfrac{v^2 + 2v - 3}{2 - v}$.   Note that
$v = -3$ and $v = 1$ are solutions of this equation.   For
$v \neq 1, -3$ separating variables gives
$\dfrac{2 - v}{(v+3)(v-1)}\, dv = \dfrac{1}{x}dx$.   Applying a partial fraction
decomposition to the left side we obtain
$[\dfrac{1}{4}\dfrac{1}{v-1} - \dfrac{5}{4}\dfrac{1}{v+3}]dv = \dfrac{dx}{x}$, and upon integrating both sides
we find that $\dfrac{1}{4}\ln|v-1| - \dfrac{5}{4}\ln|v+3| = \ln|x| + c$.
Substituting for v and performing some algebraic
manipulations we get the solution in the implicit form
$|y-x| = c|y+3x|^5$. $v = 1$ and $v = -3$ yield $y = x$ and
$y = -3x$, respectively, as solutions also.

9. $\dfrac{dy}{dx} = \dfrac{4y/x - y^3/x^3}{1 - 2y^2/x^2}$ so $x\dfrac{dv}{dx} = \dfrac{3v + v^3}{1 - 2v^2}$.   Assume $v \neq 0$ and
separate variables to get
$\dfrac{dx}{x} = \dfrac{1 - 2v^2}{v(v^2+3)}dv = (\dfrac{1/3}{v} - \dfrac{(7/3)v}{v^2+3})dv$.   Thus
$\ln|x| + c_1 = \ln|v|^{1/3} + \ln(v^2 + 3)^{-7/6}$, or
$c|x| = |y|^{1/3}x^2/(3x^2 + y^2)^{7/6}$ after simplification. $v = 0$
yields $y = 0$ as a solution also.

10. $\dfrac{dy}{dx} = \dfrac{y^4/x^4 + 2y^3/x^3 - 3y^2/x^2 - 2y/x}{2y^2/x^2 - 2y/x - 2} = \dfrac{v^4 + 2v^3 - 3v^2 - 2v}{2v^2 - 2v - 2}$
and thus $x\dfrac{dv}{dx} = \dfrac{v^4 - v^2}{2v^2 - 2v -2}$.   Assuming $v \neq 0, \pm 1$ we get
$\dfrac{2(v^2-v-1)}{v^2(v^2-1)}dv = \dfrac{dx}{x}$.   The partial fraction expansion of the
left side yields $(\dfrac{2}{v} + \dfrac{2}{v^2} - \dfrac{2v}{v^2-1})dv = \dfrac{dx}{x}$ and thus we
obtain $2\ln|v| - \dfrac{2}{v} - \ln|v^2 - 1| = \ln|x| + c_1$.   Setting
$v = y/x$ and simplifying yields $c|x| = y^2e^{-2x/y}/|y^2 - x^2|$.
We also have $y = 0$, $y = x$ and $y = -x$ as solutions from
$v = 0, \pm 1$.

15b. Making the suggested substitution we obtain

$$\frac{dY}{dX} = \frac{2(Y-k) - (X-h) + 5}{2(X-h) - (Y-k) - 4} = \frac{2Y-X+(h-2k+5)}{2X-Y+(k-2h-4)},$$ which will be

homogeneous in X, Y provided the two expressions in parenthesis are both zero. Solving these yields $h = -1$ and $k = 2$. From part(a), it then follows that $|Y-X| = c|Y+X|^3$. Substituting $Y = y + 2$, $X = x - 1$ gives the solution in the form $|y - x + 3| = c|y + x + 1|^3$.

16. The transformation $y = Y - 1$, $x = X - 3$ reduces the equation to Problem 6.

18. As in previous problems we can write the D.E. as

$$x\frac{dv}{dx} = -\frac{2v(v+2)}{v+1}.$$ Separating variables, using a partial

fraction decomposition, and solving gives $|v|^{1/2}|v+2|^{1/2}x^2 = c$. Substituting for v and simplifying yields $x^2y^2 + 2x^3y = c$ which is the same as the answer found in Example 4, Section 2.8.

19. To show that $\mu(x,y)$ is an integrating factor, multiply both terms of the D.E. by the given $\mu(x,y)$. Since the D.E. is homogeneous, we know that $M(x,y)/N(x,y) = F(y/x)$ and thus the D.E. may be written as

$$\frac{F}{xF + y}dx + \frac{1}{xF + y}dy = 0$$ after multiplying both

numerators and denominators by $1/N$ and substituting F for M/N. It may now be shown that $\frac{\partial}{\partial x}[\frac{1}{xF + y}] = \frac{\partial}{\partial y}[\frac{F}{xF + y}]$ and hence the equation is exact.

20b. As in Problem 19 we have

$$\mu(x,y) = \frac{1}{x^3 + 3xy^2 - 2xy^2} = \frac{1}{x(x^2 + y^2)}.$$ Multiplying the

D.E. by this $\mu(x,y)$ yields

$$\frac{x^2 + 3y^2}{x(x^2 + y^2)}dx - \frac{2xy}{x(x^2 + y^2)}dy = 0,$$ which can be shown to be

exact. Thus there exists a $\psi(x,y)$ such that

$$\psi_y(x,y) = -\frac{2y}{x^2 + y^2},$$ so upon integrating with respect to y

we obtain $\psi(x,y) = -\ln(x^2 + y^2) + g(x)$. We must also

have $\psi_x(x,y) = -\frac{2x}{x^2 + y^2} + g'(x) = \frac{x^2 + 3y^2}{x(x^2 + y^2)},$ so

$$g'(x) = \frac{x^2 + 3y^2 + 2x^2}{x(x^2 + y^2)} = \frac{3}{x}.$$ Thus $g(x) = 3\ln|x| + \ln c$, and

$$\psi(x,y) = -\ln(x^2 + y^2) + 3\ln|x| + \ln c, \text{ or } x^3 = c(x^2 + y^2).$$

Section 2.10, Page 94

Before trying to find the solution of a D.E. it is necessary to know its type. The student should first classify the D.E. before reading this section, which identifies the type of each equation in Problems 1 through 32.

1. Linear

2. Homogeneous

3. Exact

4. Linear equation in x

5. Exact

6. Linear

7. Linear equation in u

8. Linear

9. Exact

10. Integrating factor depends on x only

11. Exact

12. Linear

13. Homogeneous

14. Exact or homogeneous

15. Separable

16. Homogeneous

17. Linear

18. Linear or homogeneous

19. Integrating factor depends on x only

20. Separable

21. Homogeneous

22. Separable

23. Bernoulli equation

24. Separable

25. Exact

26. Integrating factor depends on x only

27. Integrating factor depends on x only

28. Exact

29. Homogeneous

30. Linear equation in x

31. Separable

32. Integrating factor depends on y only

34b. Comparing the given D.E. with the D.E. of Problem 33 we
     see that in this case $q_2(t) = -1/t$ and $q_3(t) = 1$. Hence,
     using the method suggested in Problem 33 we must solve

     $\dfrac{dv}{dt} = -(-1/t + 2y_1)v - 1$, where $y_1(t)$ is given as $1/t$.

     This is a linear first-order equation which has the

     solution $v(t) = (\dfrac{c - t^2}{2})/t$ and thus a general solution

     to the Riccati equation is $y_2(t) = 1/t + 2t/(c - t^2)$.

37b. Following the procedure outlined in Problem 36, we first
     find that the given family of curves has the slope

     $dy/dx = \dfrac{-x + c}{y}$ at each point. To eliminate c, solve the

     given equation for c to get $c = \dfrac{x^2 + y^2}{2x}$. Thus

     $dy/dx = \dfrac{y^2 - x^2}{2xy}$ and the orthogonal family of curves will

     then have slopes given by $dy/dx = \dfrac{-2xy}{y^2 - x^2}$. This equation

     may be written as $y' = \dfrac{-2(y/x)}{(y/x)^2 - 1}$ so it is homogeneous.

     Substitution of $y = vx$, separation of variables, and a
     partial fraction decomposition lead to the equation

     $(\dfrac{1}{v} - \dfrac{2v}{1+v^2})dv = \dfrac{dx}{x}$ whose solution is given implicitly by

     $\dfrac{v}{1 + v^2} = \hat{c}x$. Substitution for v and algebraic

     simplification give $\dfrac{y}{x^2+y^2} = \hat{c}$ or $x^2 + y^2 = 2cy$.

     Completion of the square gives the answer in more easily
     recognizable form.

37d. The solution $y + \sqrt{x^2+y^2} = \tilde{c}\,x^2$ is obtained if the
     procedure used in Problem 37b is followed. This answer
     can be put in the desired (and preferable) form by
     subtracting y from both sides and then squaring to obtain
     $x^2+y^2 = (\tilde{c}x^2-y)^2 = \tilde{c}^2x^4 - 2\tilde{c}x^2y + y^2$. Subtracting y^2 and
     division by x^2 ($x \neq 0$) yields the answer.

38b. The slope of the given family of curves is given by
     $m_2 = dy/dx = -x/y$ and $\tan\theta = 1$. Thus, from the given

relationship of θ, m_1 and m_2 we have $1 = \dfrac{m_1 + x/y}{1 - m_1(x/y)}$.

Solving for m_1, we find the intersecting family of curves must have slopes $m_1 = dy/dx = (y-x)/(y+x)$, which is a homogeneous D.E.

39. Using the equation for the slope of a line passing through the points (x,y) and (a,b) we find the equation to solve is $y' = (y-b)/(x-a)$, which is separable.

41. In this case the point through which the tangent line must pass is $(0,y/2)$ and the equation to solve is $y' = y/2x$.

42. If $S(t)$ is the value of the business, then the D.E. is $S' = kS^2$, subject to the conditions $S(0) = 1$ million, $S(1) = 2$ million. Separating variables and integrating leads to $S(t) = -\dfrac{1}{kt + c}$. The I.C. $S(0) = 1$ gives $c = -1$. Using the second condition, $S(1) = 2 = -\dfrac{1}{k - 1}$ gives $k = \dfrac{1}{2}$ so the solution is $S(t) = \dfrac{2}{2 - t}$ million. In 6 months, $t = \dfrac{3}{2}$ (since $t = 0$ occurred 1 year ago) and $S(\dfrac{3}{2}) = 4$ million. As t approaches 2, his wealth approaches infinity, so the solution cannot be extended until two years from now (which would be $t = 3$).

43a. The D.E. is $dV/dt = k - \alpha \pi r^2$. The volume of a cone of height L and radius r is given by $V = \pi r^2 L/3$ where $L = hr/a$ from symmetry. Solving for r yields the desired solution.

43b. From the material on logistic growth in Section 2.6 we have equilibrium given by $k - \alpha \pi r^2 = 0$.

Section 2.11, Page 105

1. Let $s = t-1$ and $w(s) = y(t(s)) - 2$, then when $t = 1$ and $y = 2$ we have $s = 0$ and $w(0) = 0$. Also, $\dfrac{dw}{ds} = \dfrac{dw}{dt} \cdot \dfrac{dt}{ds} = \dfrac{d}{dt}(y-2) \dfrac{dt}{ds} = \dfrac{dy}{dt}$ and hence

$$\frac{dw}{ds} = (s+1)^2 + (w+2)^2,$$ upon substitution into the given D.E.

7. Using Eq. (7) and following the steps of the Example of this section we obtain the following:

$$\phi_1(t) = \int_0^t (s\phi_0(s) + 1)\,ds = s\Big|_0^t = t$$

$$\phi_2(t) = \int_0^t (s^2 + 1)\,ds = \left(\frac{s^3}{3} + s\right)\Big|_0^t = t + \frac{t^3}{3}$$

$$\phi_3(t) = \int_0^t \left(s^2 + \frac{s^4}{3} + 1\right)ds = \left(\frac{s^3}{3} + \frac{s^5}{3\cdot5} + s\right)\Big|_0^t = t + \frac{t^3}{3} + \frac{t^5}{3\cdot5}.$$

Based upon these we hypothesize that:

$$\phi_n(t) = \sum_{k=1}^{n} \frac{t^{2k-1}}{1\cdot3\cdot5\cdots(2k-1)}$$ and use mathematical induction to

verify this form for $\phi_n(t)$. Using Eq. (7) again we have:

$$\phi_{n+1}(t) = \int_0^t \left(\sum_{k=1}^{n} \frac{s^{2k}}{1\cdot3\cdot5\cdots(2k-1)} + 1\right)ds$$

$$= \sum_{k=1}^{n} \frac{t^{2k+1}}{1\cdot3\cdot5\cdots(2k+1)} + t$$

$$= \sum_{k=0}^{n} \frac{t^{2k+1}}{1\cdot3\cdot5\cdots(2k+1)}$$

$$= \sum_{i=1}^{n+1} \frac{t^{2i-1}}{1\cdot3\cdot5\cdots(2i-1)},$$ where $i = k+1$. Since this is

the same form for $\phi_{n+1}(t)$ as derived from $\phi_n(t)$ above, we have verified by mathematical induction that $\phi_n(t)$ is as given.

Section 2.12, Page 117

2. Using the given difference equation we have for $n=0$, $y_1 = y_0/2$; for $n=1$, $y_2 = 2y_1/3 = y_0/3$; and for $n=2$, $y_3 = 3y_2/4 = y_0/4$. Thus we guess that $y_n = y_0/(n+1)$, and the given equation then gives $y_{n+1} = \frac{n+1}{n+2}y_n = y_0/(n+2)$, which, by mathematical induction, verifies $y_n = y_0/(n+1)$ as the solution for all n.

5. From the given equation we have $y_1 = .5y_0+6$.

$y_2 = .5y_1 + 6 = (.5)^2 y_0 + 6(1 + \frac{1}{2})$ and

$y_3 = .5y_2 + 6 = (.5)^3 y_0 + 6(1 + \frac{1}{2} + \frac{1}{4})$. In general, then

$y_n = (.5)^n y_0 + 6(1 + \frac{1}{2} + \cdots + \frac{1}{2^{n-1}})$

$\quad = (.5)^n y_0 + 6(\frac{1 - (1/2)^n}{1 - 1/2})$

$\quad = (.5)^n y_0 + 12 - (.5)^n 12$

$\quad = (.5)^n (y_0 - 12) + 12$. Mathematical induction can now be
used to prove that this is the correct solution.

10. The governing equation is $y_{n+1} = \rho y_n - b$, which has the

solution $y_n = \rho^n y_0 - \frac{1-\rho^n}{1-\rho} b$ (Eq.(14) with a negative b).
Setting $y_{360} = 0$ and solving for b we obtain

$b = \frac{(1-\rho)\rho^{360} y_0}{1-\rho^{360}}$, where $\rho = 1.0075$ for part a.

13. You must solve Eq.(14) numerically for ρ when $y_{240} = 0$,
$b = -\$900$ and $y_0 = \$95,000$.

14. Substituting $u_n = \frac{\rho-1}{\rho} + v_n$ into Eq.(21) we get

$\frac{\rho-1}{\rho} + v_{n+1} = \rho(\frac{\rho-1}{\rho} + v_n)(1 - \frac{\rho-1}{\rho} - v_n)$ or

$v_{n+1} = -\frac{\rho-1}{\rho} + (\rho-1 + \rho v_n)(\frac{1}{\rho} - v_n)$

$\quad = \frac{1-\rho}{\rho} + \frac{\rho-1}{\rho} - (\rho-1)v_n + v_n - \rho v_n^2 = (2-\rho)v_n - \rho v_n^2$.

15a. For $u_0 = .2$ we have $u_1 = 3.2u_0(1-u_0) = .512$ and
$u_2 = 3.2u_1(1-u_1) = .7995392$. Likewise $u_3 = .51288406$,
$u_4 = .7994688$, $u_5 = .51301899$, $u_6 = .7994576$ and
$u_7 = .5130404$. Continuing in this fashion,
$u_{14} = u_{16} = .79945549$ and $u_{15} = u_{17} = .51304451$.

17. For both parts of this problem a computer spreadsheet
was used and an initial value of $u_0 = .2$ was chosen.
Different initial values or different computer programs
may need a slightly different numbe of iterations to
reach the limiting value.

17a. The limiting value of .65517 (to 5 decimal places) is
     reached after approximately 100 iterations for ρ = 2.9.
     The limiting value of .66102 (to 5 decimal places) is
     reached after approximately 200 iterations for ρ = 2.95.
     The limiting value of .66555 (to 5 decimal places) is
     reached after approximately 910 iterations for ρ = 2.99.

17b. The solution oscillates between .63285 and .69938 after
     approximately 400 iterations for ρ = 3.01. The solution
     oscillates between .59016 and .73770 after approximately
     130 iterations for ρ = 3.05. The solution oscillates
     between .55801 and .76457 after approximately 30
     iterations for ρ = 3.1. For each of these cases
     additional iterations verified the oscillations were
     correct to five decimal places.

18. For an initial value of .2 and ρ = 3.448 we have the
     solution oscillating between .4403086 and .8497146.
     After approximately 3570 iterations the eighth decimal
     place is still not fixed, though. For the same initial
     value and ρ = 3.45 the solution oscillates between the
     four values: .43399155, .84746795, .44596778 and
     .85242779 after 3700 iterations.. For ρ = 3.449, the
     solution is still varying in the fourth decimal place
     after 3570 iterations, but there appear to be four
     values.

CHAPTER 3

Section 3.1, Page 128

1. Assume $y = e^{rt}$, which is substituted into the D.E. to
   obtain the characteristic equation $r^2 + 2r - 3 = 0$.
   Hence $r_1 = 1$ and $r_2 = -3$ and the general solution is then
   $y = c_1e^t + c_2e^{-3t}$.

5. The characteristic equation is $r^2 + 5r = 0$, so the roots
   are $r_1 = 0$, and $r_2 = -5$. Thus
   $y = c_1e^{ot} + c_2e^{-5t} = c_1 + c_2e^{-5t}$.

7. The characteristic equation is $r^2 - 9r + 9 = 0$ so that
   $r = (9\pm\sqrt{81-36})/2 = (9\pm3\sqrt{5})/2$. Hence
   $y = c_1\exp[(9+3\sqrt{5})t/2] + c_2\exp[(9-3\sqrt{5})t/2]$.

10. Substituting $y = e^{rt}$ in the D.E.
    we obtain the characteristic
    equation $r^2 + 4r + 3 = 0$, which
    has the roots $r_1 = -1$, $r_2 = -3$.
    Thus $y = c_1e^{-t} + c_2e^{-3t}$ and
    $y' = -c_1e^{-t} - 3c_2e^{-3t}$.
    Substituting $t = 0$ we then have
    $c_1 + c_2 = 2$ and $-c_1 - 3c_2 = -1$,
    yielding $c_1 = 5/2$ and

    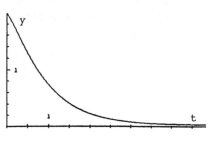

    $c_2 = -1/2$. Thus $y = \dfrac{5}{2}e^{-t} - \dfrac{1}{2}e^{-3t}$
    and hence $y \to 0$ as $t \to \infty$.

15. The characteristic equation is $r^2 + 8r - 9 = 0$, so that
    $r_1 = 1$ and $r_2 = -9$ and the general solution is
    $y = c_1e^t + c_2e^{-9t}$. Since the I.C. are given at $t = 1$, it
    is convenient to write the general solution in the form
    $y = k_1e^{(t-1)} + k_2e^{-9(t-1)}$. Note that
    $c_1 = k_1e^{-1}$ and $c_2 = k_2e^9$. The advantage of the latter
    form of the general solution becomes clear when we apply
    the I.C. $y(1) = 1$ and $y'(1) = 0$. This latter form of y

gives $y' = k_1 e^{(t-1)} - 9k_2 e^{-9(t-1)}$ and thus setting $t = 1$ in y and y' yields the equations $k_1 + k_2 = 1$ and $k_1 - 9k_2 = 0$. Solving for k_1 and k_2 we find that

$$y = (9e^{(t-1)} + e^{-9(t-1)})/10.$$

17. Comparing the given solution to Eq(17), we see that $r_1 = 2$ and $r_2 = -3$ are the two roots of the characteristic equation. Thus we have $(r-2)(r+3) = 0$, or $r^2 + r - 6 = 0$ as the characteristic equation. Hence the given solution is for the D.E. $y'' + y' - 6y = 0$.

19. The roots of the characteristic equation are $r = 1, -1$ and thus the general solution is $y(t) = c_1 e^t + c_2 e^{-t}$.

$y(0) = c_1 + c_2 = \dfrac{5}{4}$ and $y'(0) = c_1 - c_2 = -\dfrac{3}{4}$, yielding

$y(t) = \dfrac{1}{4}e^t + e^{-t}$. From this $y'(t) = \dfrac{1}{4}e^t - e^{-t} = 0$ or

$e^{2t} = 4$ or $t = \ln 2$. The second derivative test or a graph of the solution indicates this is a minimum point.

21. The general solution is $y = c_1 e^{-t} + c_2 e^{2t}$. Using the I.C. we obtain $c_1 + c_2 = \alpha$ and $-c_1 + 2c_2 = 2$, so adding the two equations we find $3c_2 = \alpha + 2$. If y is to approach zero as $t \to \infty$, c_2 must be zero. Thus $\alpha = -2$.

24. The roots of the characteristic equation are given by $r = -2, \alpha - 1$ and thus $y(t) = c_1 e^{-2t} + c_2 e^{(\alpha-1)t}$. Hence, for $\alpha < -1$, all solutions tend to zero as $t \to \infty$. For $\alpha > -1$, the second term becomes unbounded, but not the first, so there are no values of α for which all solutions become unbounded.

25c. From part (a), if $\beta = 2$ then $y(t) = e^{-2t}$ and the solution simply decays to zero. For $\beta > 2$, the solution becomes unbounded negatively, and again there is no minimum point.

27. The second solution must decay faster than e^{-t}, so choose e^{-2t} or e^{-3t} etc. as the second solution. Then proceed as

in Problem 17.

28. Let $v = y'$, then $v' = y''$ and thus the D.E. becomes
$t^2 v' + 2tv - 1 = 0$ or $t^2 v' + 2tv = 1$. The left side is
recognized as $(t^2 v)'$ and thus we may integrate to obtain
$t^2 v = t + c$ (otherwise, divide both sides of the D.E. by
t^2 and find the integrating factor, which is just t^2 in
this case). Solving for $v = dy/dt$ we find
$dy/dt = 1/t + c/t^2$ so that $y = \ln t + c_1/t + c_2$

30. Set $v = y'$, then $v' = y''$ and thus the D.E. becomes
$v' + tv^2 = 0$. This equation is separable and has the
solution $-v^{-1} + t^2/2 = c$ or $v = y' = -2/(c_1 - t^2)$ where
$c_1 = 2c$. We must consider separately the cases $c_1 = 0$,
$c_1 > 0$ and $c_1 < 0$. If $c_1 = 0$, then $y' = 2/t^2$ or
$y = -2/t + c_2$. If $c_1 > 0$, let $c_1 = k^2$. Then
$y' = -2/(k^2-t^2) = -(1/k)[1/(k-t) + 1/(k+t)]$, so that
$y = (1/k)\ln|(k-t)/(k+t)|+c_2$. If $c_1 < 0$, let $c_1 = -k^2$.
Then $y' = 2/(k^2 + t^2)$ so that $y = (2/k)\tan^{-1}(t/k) + c_2$.
Finally, we note that $y = $ constant is also a solution of
the D.E.

34. Following the procedure outlined, let $v = dy/dt$ and
$y'' = dv/dt = v\,dv/dy$. Thus the D.E. becomes
$yv\,dv/dy + v^2 = 0$, which is a separable equation with the
solution $v = c_1/y$. Next let $v = dy/dt = c/y$, which again
separates to give the solution $y^2 = c_1 t + c_2$.

37. Again let $v = y'$ and $v' = v\,dv/dy$ to obtain
$2y^2 v\,dv/dy + 2yv^2 = 1$. This is an exact equation with
solution $v = \pm y^{-1}(y + c_1)^{1/2}$. To solve this equation,
we write it in the form $\pm y\,dy/(y+c_1)^{1/2} = dt$. On
observing that the left side of the equation can be
written as $\pm[(y+c_1) - c_1]dy/(y+c_1)^{1/2}$ we integrate and
find $\pm (2/3)(y-2c_1)(y+c_1)^{1/2} = t + c_2$.

39. If $v = y'$, then $v' = v\,dv/dy$ and the D.E. becomes $v\,dv/dy + v^2 = 2e^{-y}$. Dividing by v we obtain $dv/dy + v = 2v^{-1}e^{-y}$, which is a Bernoulli equation. Let $w(y) = v^2$, then $dw/dy = 2v\,dv/dy$ and the last D.E. becomes $dw/dy + 2w = 4e^{-y}$, which is linear in w. It's solution is $w = v^2 = ce^{-2y} + 4e^{-y}$. Setting $v = dy/dt$, we obtain a separable equation in y and t, which is solved to yield the solution.

40. Since both t and y are missing, either approach used above will work. In this case it's easier to use the approach of Problems 28-33, so let $v = y'$ and thus $v' = y''$ and the D.E. becomes $v\,dv/dt = 2$.

43. The variable y is missing. Let $v = y'$, then $v' = y''$ and the D.E. becomes $vv' - t = 0$. The solution of the separable equation is $v^2 = t^2 + c_1$. Substituting $v = y'$ and applying the I.C. $y'(1) = 1$, we obtain $y' = t$. The positive square root was chosen because $y' > 0$ at $t = 1$. Solving this last equation and applying the I.C. $y(1) = 2$, we obtain $y = t^2/2 + 3/2$.

Section 3.2, Page 138

4. $W(x, xe^x) = \begin{vmatrix} x & xe^x \\ 1 & e^x + xe^x \end{vmatrix} = xe^x + x^2e^x - xe^x = x^2e^x.$

8. As in Example 1, $p(t) = -3t/(t-1)$ and $q(t) = 4/(t-1)$, so the only point of discontinuity is $t = 1$. By Theorem 3.2.1, the largest interval is $-\infty < t < 1$, since the initial point is $t_0 = -2$.

14. For $y = t^{1/2}$, $y' = \frac{1}{2}t^{-1/2}$ and $y'' = -\frac{1}{4}t^{-3/2}$. Thus $yy'' + (y')^2 = -\frac{1}{4}t^{-1} + \frac{1}{4}t^{-1} = 0$. Similarly $y = 1$ is also a solution. If $y = c_1(1) + c_2t^{1/2}$ is substituted in the D.E. you will get $-c_1c_2/4t^{3/2}$, which is zero only if $c_1 = 0$ or $c_2 = 0$. Thus the linear combination of two solutions is not, in general, a solution. Theorem 3.2.2 is not contradicted however, since the D.E. is not linear.

15. $y = \phi(t)$ is a solution of the D.E. so $L[\phi](t) = g(t)$.
Since L is a linear operator,
$L[c\phi](t) = cL[\phi](t) = cg(t)$. But, since $g(t) \neq 0$,
$cg(t) = g(t)$ if and only if $c = 1$. This is not a
contradiction of Theorem 3.2.2 since the linear D.E. is
not homogeneous.

18. $W(t,g) = \begin{vmatrix} t & g \\ 1 & g' \end{vmatrix} = tg' - g = t^2 e^t$, or $g' - \frac{1}{t}g = te^t$. This

has an integrating factor of $\frac{1}{t}$ and thus $\frac{1}{t}g' - \frac{1}{t^2}g = e^t$

or $(\frac{1}{t}g)' = e^t$. Integrating and multiplying by t we

obtain $g(t) = te^t + ct$.

21. From Section 3.1, e^t and e^{-2t} are two solutions, and
since $W(e^t, e^{-2t}) \neq 0$ they form a fundamental set of
solutions. To find the fundamental set specified by
Theorem 3.2.5, let $y(t) = c_1 e^t + c_2 e^{-2t}$, where c_1 and c_2
satisfy
$c_1 + c_2 = 1$ and $c_1 - 2c_2 = 0$ for y_1. Solving, we find
$y_1 = \frac{2}{3}e^t + \frac{1}{3}e^{-2t}$. Likewise, c_1 and c_2 satisfy
$c_1 + c_2 = 0$ and $c_1 - 2c_2 = 1$ for y_2, so that
$y_2 = \frac{1}{3}e^t - \frac{1}{3}e^{-2t}$.

25. For $y_1 = x$, we have $x^2(0) - x(x+2)(1) + (x+2)(x) = 0$ and
for $y_2 = xe^x$ we have $x^2(x+2)e^x - x(x+2)(x+1)e^x + (x+2)xe^x = 0$.
From Problem 4, $W(x, xe^x) = x^2 e^x \neq 0$ for $x > 0$, so y_1 and
y_2 form a fundamental set of solutions.

27. Suppose that
$P(x)y'' + Q(x)y' + R(x)y = [P(x)']' + [f(x)y]'$. On
expanding the right side and equating coefficients, we
find $f'(x) = R(x)$ and $P'(x) + f(x) = Q(x)$. These two
conditions on f can be satisfied if
$R(x) = Q'(x) - P''(x)$ which gives the necessary condition
$P''(x) - Q'(x) + R(x) = 0$.

30. We have $P(x) = x$, $Q(x) = -\cos x$, and $R(x) = \sin x$ and the condition for exactness is satisfied. Also, from Problem 27, $f(x) = Q(x) - P'(x) = -\cos x - 1$, so the D.E. becomes $(xy')' - [(1 + \cos x)y]' = 0$. Hence $xy' - (1 + \cos x)y = c_1$. This is a first order linear D.E. and the integrating factor (after dividing by x) is

$\mu(x) = \exp[-\int x^{-1}(1 + \cos x)dx]$. The general solution is

$$y = [\mu(x)]^{-1}[c_1\int_{x_0}^{x} t^{-1}\mu(t)dt + c_2].$$

32. We want to choose $\mu(x)$ and $f(x)$ so that $\mu(x)P(x)y'' + \mu(x)Q(x)y' + \mu(x)R(x)y = [\mu(x)P(x)y']' + [f(x)y]'$. Expand the right side and equate coefficients of y'', y' and y. This gives $\mu'(x)P(x) + \mu(x)P'(x) + f(x) = \mu(x)Q(x)$ and $f'(x) = \mu(x)R(x)$. Differentiate the first equation and then eliminate $f'(x)$ to obtain the adjoint equation $P\mu'' + (2P' - Q)\mu' + (P'' - Q' + R)\mu = 0$.

36. Write the adjoint D.E. given in Problem 32 as $\hat{P}\mu'' + \hat{Q}\mu' + \hat{R}\mu = 0$ where $\hat{P} = P$, $\hat{Q} = 2P' - Q$, and $\hat{R} = P'' - Q' + R$. The adjoint of this equation, namely the adjoint of the adjoint, is $\hat{P}y'' + (2\hat{P}' - \hat{Q})y' + (\hat{P}'' - \hat{Q}' + \hat{R})y = 0$. After substituting for $\hat{P}$, $\hat{Q}$, and $\hat{R}$ and simplifying, we obtain $Py'' + Qy' + Ry = 0$. This is the same as the original equation.

37. From Problem 32 the adjoint of $Py'' + Qy' + Ry = 0$ is $P\mu'' + (2P' - Q)\mu' + (P'' - Q' + R)\mu = 0$. The two equations are the same if $2P' - Q = Q$ and $P'' - Q' + R = R$. This will be true if $P' = Q$. Hence the original D.E. is self-adjoint if $P' = Q$. For Problem 33, $P(x) = x^2$ so $P'(x) = 2x$ and $Q(x) = x$. Hence the Bessel equation of order ν is not self-adjoint. In a similar manner we find that Problems 34 and 35 are self-adjoint.

Section 3.3, Page 144

2. Since $\cos 3\theta = 4\cos^3\theta - 3\cos\theta$ we have $\cos 3\theta - (4\cos^3\theta - 3\cos\theta) = 0$ for all θ. From Eq.(1) we have $k_1 = 1$ and $k_2 = -1$ and thus $\cos 3\theta$ and $4\cos^3\theta - 3\cos\theta$ are linearly dependent.

6. $W(t,t^{-1}) = -2/t \neq 0$.

7.   For $t>0$ $g(t) = t$ and hence $f(t) - 3g(t) = 0$ for all t.
     Therefore f and g are linearly dependent on $0<t$. For $t<0$
     $g(t) = -t$ and $f(t) + 3g(t) = 0$, so again f and g are
     linearly dependent on $t<0$. For any interval that
     includes the origin, such as $-1<t<2$, there is no c for
     which $f(t) + cg(t) = 0$ for all t, and hence f and g are
     linearly independent on this interval.

12. The D.E. is linear and homogeneous. Hence, if y_1 and y_2
     are solutions, then $y_3 = y_1 + y_2$ and $y_4 = y_1 - y_2$ are
     solutions. $W(y_3,y_4) = y_3 y_4' - y_3' y_4 = (y_1 + y_2)(y_1' - y_2') -$
     $(y_1' + y_2')(y_1 - y_2) = -2(y_1 y_2' - y_1' y_2) = -2W(y_1,y_2)$, is not
     zero since y_1 and y_2 are linearly independent solutions.
     Hence y_3 and y_4 form a fundamental set of solutions.

15. Writing the D.E. in the form of Eq.(6), we have
     $p(t) = -(t+2)/t$. Thus Eq.(7) yields
     $$W(t) = cexp[-\int \frac{-(t+2)}{t} dt] = ct^2 e^t.$$

20. From Eq.(7) we have $W(y_1,y_2) = cexp[-\int p(t)dt]$, where
     $p(t) = 2/t$ from the D.E. Thus $W(y_1,y_2) = c/t^2$. Since
     $W(y_1,y_2)(1) = 2$ we find $c = 2$ and thus $W(y_1,y_2)(5) = 2/25$.

24. Let c be in the point in I at which both y_1 and y_2
     vanish. Then $W(y_1,y_2)(c) = y_1(c)y_2'(c) - y_1'(c)y_2(c) = 0$.
     Hence, by Theorem 3.3.3 the functions y_1 and y_2 cannot
     form a fundamental set.

26. Suppose that y_1 and y_2 have a point of inflection at t_0
     and either $p(t_0) \neq 0$ or $q(t_0) \neq 0$. Since $y_1''(t_0) = 0$ and
     $y_2''(t_0) = 0$ it follows from the D.E. that $p(t_0)y_1'(t_0) +$
     $q(t_0)y_1(t_0) = 0$ and $p(t_0)y_2'(t_0) + q(t_0)y_2(t_0) = 0$. If
     $p(t_0) = 0$ and $q(t_0) \neq 0$ then $y_1(t_0) = y_2(t_0) = 0$, and
     $W(y_1,y_2)(t_0) = 0$ so the solutions cannot form a
     fundamental set. If $p(t_0) \neq 0$ and $q(t_0) = 0$ then

$y_1'(t_0) = y_2'(t_0) = 0$ and $W(y_1, y_2)(t_0) = 0$, so again the solutions cannot form a fundamental set. If $p(t_0) \neq 0$ and $q(t_0) \neq 0$ then $y_1'(t_0) = q(t_0)y_1(t_0)/p(t_0)$ and

$y_2'(t_0) = q(t_0)y_2(t_0)/p(t_0)$ and thus

$$W(y_1, y_2)(t_0) = y_1(t_0)y_2'(t_0) - y_1'(t_0)y_2(t_0)$$
$$= y_1(t_0)[q(t_0)y_2(t_0)/p(t_0)] -$$
$$[q(t_0)y_1(t_0)/p(t_0)]y_2(t_0)$$
$$= 0.$$

27. Let $-1 < t_0, t_1 < 1$ and $t_0 \neq t_1$. If $y_1 = t$ and $y_2 = t^2$ are linearly dependent then $c_1 t_1 + c_2 t_1^2 = 0$ and

$c_1 t_0 + c_2 t_0^2 = 0$ have a solution for c_1 and c_2 such that c_1 and c_2 are not both zero. But this system of equations has a non-zero solution only if $t_1 = 0$ or $t_0 = 0$ or $t_1 = t_0$. Hence, the only set c_1 and c_2 that satisfies the system for every choice of t_0 and t_1 in $-1 < t < 1$ is

$c_1 = c_2 = 0$. Therefore t and t^2 are linearly independent on $-1 < t < 1$. Next, $W(t, t^2) = t^2$ clearly vanishes at $t = 0$. Since $W(t, t^2)$ vanishes at $t = 0$, but t and t^2 are linearly independent on $-1 < t < 1$, it follows that t and t^2 cannot be solutions of Eq.(6) on $-1 < t < 1$. To show that the functions $y_1 = t$ and $y_2 = t^2$ are solutions of

$t^2 y'' - 2ty'' + 2y = 0$, substitute each of them in the equation. Clearly, they are solutions. There is no contradiction to Theorem 3.3.2 since $p(t) = -2/t$ and $q(t) = 2/t^2$ are discontinuous at $t = 0$, and hence the theorem does not apply on the interval $-1 < t < 1$.

28. On $0 < t < 1$, $f(t) = t^3$ and $g(t) = t^3$. Hence there are nonzero constants, $c_1 = 1$ and $c_2 = -1$, such that $c_1 f(t) + c_2 g(t) = 0$ for each t in $(0,1)$. On $-1 < t < 0$,

$f(t) = -t^3$ and $g(t) = t^3$; thus $c_1 = c_2 = 1$ defines constants such that $c_1f(t) + c_2g(t) = 0$ for each t in $(-1,0)$. Thus f and g are linearly dependent on $0 < t < 1$ and on $-1 < t < 0$. We will show that $f(t)$ and $g(t)$ are linearly independent on $-1 < t < 1$ by demonstrating that it is impossible to find constants c_1 and c_2, not both zero, such that $c_1f(t) + c_2g(t) = 0$ for all t in $(-1,1)$. Assume that there are two such nonzero constants and choose two points t_0 and t_1 in $-1 < t < 1$ such that $t_0 < 0$ and $t_1 > 0$. Then $-c_1t_0^3 + c_2t_0^3 = 0$ and $c_1t_1^3 + c_2t_1^3 = 0$. These equations have a nontrivial solution for c_1 and c_2 only if the determinant of coefficients is zero. But the determinant of coefficients is $-2t_0^3t_1^3 \neq 0$ for t_0 and t_1 as specified. Hence $f(t)$ and $g(t)$ are linearly independent on $-1 < t < 1$.

Section 3.4, Page 150

1. $\exp(1+2i) = e^{1+2i} = ee^{2i} = e(\cos 2 + i\sin 2)$.

5. Recall that $2^{1-i} = e^{\ln(2^{1-i})} = e^{(1-i)\ln 2}$.

7. As in Section 3.1, we seek solutions of the form $y = e^{rt}$. Substituting this into the D.E. yields the characteristic equation $r^2 - 2r + 2 = 0$, which has the roots $r_1 = 1 + i$ and $r_2 = 1 - i$, using the quadratic formula. Thus $\lambda = 1$ and $\mu = 1$ and from Eq.(13) the general solution is $y = c_1e^t\cos t + c_2e^t\sin t$.

14. The characteristic equation is $9r^2 + 9r - 4$, which has the real roots $-4/3$ and $1/3$. Thus the solution has the same form as in Section 3.1, $y(t) = c_1e^{t/3} + c_2e^{-4t/3}$.

18. The characteristic equation is
$r^2 + 4r + 5 = 0$, which has the
roots $r_1, r_2 = -2 \pm i$.   Thus

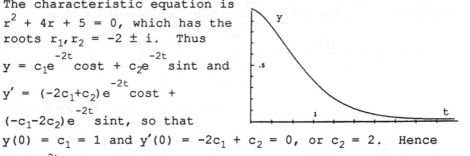

$y = c_1 e^{-2t} \cos t + c_2 e^{-2t} \sin t$ and

$y' = (-2c_1 + c_2) e^{-2t} \cos t +$

$(-c_1 - 2c_2) e^{-2t} \sin t$, so that

$y(0) = c_1 = 1$ and $y'(0) = -2c_1 + c_2 = 0$, or $c_2 = 2$.   Hence

$y = e^{-2t} (\cos t + 2\sin t)$.

22. The characteristic equation is
$r^2 + 2r + 2 = 0$, so $r_1, r_2 = -1 \pm i$.
Since the I.C. are given at $\pi/4$
we assume $y = e^{-(t-\pi/4)} (c_1 \cos t + c_2 \sin t)$

so $y' = -e^{-(t-\pi/4)} (c_1 \cos t + c_2 \sin t) +$

$e^{-(t-\pi/4)} (-c_1 \sin t + c_2 \cos t)$.   Thus $\sqrt{2}\, c_1/2 + \sqrt{2}\, c_2/2 = 2$

and $-\sqrt{2}\, c_1 = -2$ and hence $y = \sqrt{2}\, e^{-(t-\pi/4)} (\cos t + \sin t)$.

23a. The characteristic equation is $3r^2 - r + 2 = 0$, which has

the roots $r_1, r_2 = \dfrac{1}{6} \pm \dfrac{\sqrt{23}}{6} i$.   Thus $u(t) =$

$e^{t/6} (c_1 \cos \dfrac{\sqrt{23}}{6} t + c_2 \sin \dfrac{\sqrt{23}}{6} t)$ and we obtain $u(0) = c_1 = 2$

and $u'(0) = \dfrac{1}{6} c_1 + \dfrac{\sqrt{23}}{6} c_2 = 0$.   Solving for c_1 and c_2 we

find $u(t) = e^{t/6} (2\cos \dfrac{\sqrt{23}}{6} t - \dfrac{2}{\sqrt{23}} \sin \dfrac{\sqrt{23}}{6} t)$.

23b. To estimate the first time that $|u(t)| = 10$ plot the
graph of $u(t)$ as found in part (a).   Use this estimate in
an appropriate computer software program to find
$t = 10.7598$.

25a. The characteristic equation is $r^2 + 2r + 6 = 0$, so

$r_1, r_2 = -1 \pm \sqrt{5}\, i$ and $y(t) = e^{-t} (c_1 \cos\sqrt{5}\, t + c_2 \sin\sqrt{5}\, t)$.

Thus $y(0) = c_1$ and $y'(0) = -c_1 + \sqrt{5}\, c_2 = \alpha$ and hence $y(t)$

$= e^{-t} (2\cos\sqrt{5}\, t + \dfrac{\alpha+2}{\sqrt{5}} \sin \sqrt{5}\, t)$.

25b. $y(1) = e^{-1}(2\cos\sqrt{5} + \dfrac{\alpha+2}{\sqrt{5}}\sin\sqrt{5}) = 0$ and hence

$$\alpha = -2 - \frac{2\sqrt{5}}{\tan\sqrt{5}} = 1.50878.$$

25c. For $y(t) = 0$ we must have $2\cos\sqrt{5}\,t + \dfrac{\alpha+2}{\sqrt{5}}\sin\sqrt{5}\,t = 0$ or

$$\tan\sqrt{5}t = \frac{-2\sqrt{5}}{\alpha+2}\cdot.$$ For $\alpha > -2$ this yields

$$\sqrt{5}\,t = \pi - \arctan\frac{2\sqrt{5}}{\alpha+2}$$ since $\arctan x < 0$ when $x < 0$.

25d. From part (c) $\arctan\dfrac{2\sqrt{5}}{\alpha+2} \to 0$ as $\alpha \to \infty$, so $t \to \pi/\sqrt{5}$.

31. $\dfrac{d}{dt}[e^{\lambda t}(\cos\mu t + i\sin\mu t)] = \lambda e^{\lambda t}(\cos\mu t + i\sin\mu t)$

$+\ e^{\lambda t}(-\mu\sin\mu t + i\mu\cos\mu t) = \lambda e^{\lambda t}(\cos\mu t + i\sin\mu t)$

$+\ i\mu e^{\lambda t}(i\sin\mu t + \cos\mu t) = e^{\lambda t}(\lambda+i\mu)(\cos\mu t + i\sin\mu t).$

Setting $r = \lambda+i\mu$ we then have $\dfrac{d}{dt}e^{rt} = re^{rt}$.

33. Suppose that $t = a$ and $t = b$ $(b>a)$ are consecutive zeros
of y_1. We must show that y_2 vanishes once and only once
in the interval $a < t < b$. Assume that it does not
vanish. Then we can form the quotient y_1/y_2 on the
interval $a \le t \le b$. Note $y_2(a) \ne 0$ and $y_2(b) \ne 0$,
otherwise y_1 and y_2 would not be linearly independent
solutions. Next, y_1/y_2 vanishes at $t = a$ and $t = b$ and
has a derivative in $a < t < b$. By Rolles theorem, the
derivative must vanish at an interior point. But
$(\dfrac{y_1}{y_2})' = \dfrac{y_1'y_2 - y_2'y_1}{y_2^{\,2}} = \dfrac{-W(y_1,y_2)}{y_2^{\,2}}$, which cannot be zero
since y_1 and y_2 are linearly independent solutions.
Hence we have a contradiction, and we conclude that y_2
must vanish at a point between a and b. Finally, we show
that it can vanish at only one point between a and b.
Suppose that it vanishes at two points c and d between a
and b. By the argument we have just given we can show
that y_1 must vanish between c and d. But this contradicts
the hypothesis that a and b are consecutive zeros of y_1.

35. We use the result of Problem 34. Note that $q(t) = e^{-t^2} > 0$
for $-\infty < t < \infty$. Next, we find that $(q' + 2pq)/q^{3/2} = 0$.
Hence the D.E. can be transformed into an equation with
constant coefficients by letting $x = u(t) = \int e^{-t^2/2}\, dt$.
Substituting $x = u(t)$ in the differential equation found
in part (b) of Problem 34 we obtain, after dividing by
the coefficient of d^2y/dx^2, the D.E. $d^2y/dx^2 - y = 0$.
Hence the general solution of the original D.E. is
$$y = c_1\cos x + c_2\sin x, \quad x = \int e^{-t^2/2}\, dt.$$

38. Rewrite the D.E. as $y'' + (\alpha/t)y' + (\beta/t^2)y = 0$ so that
$p = \alpha/t$ and $q = \beta/t^2$, which satisfy the conditions of
parts (c) and (d) of Problem 34. Thus
$x = \int (1/t^2)^{1/2}dt = \ln t$ will transform the D.E. into
$dy^2/dx^2 + (\alpha-1)\, dy/dx + \beta y = 0$. Note that since β is
constant, it can be neglected in defining x.

39. By direct substitution or from Problem 38, $x = \ln t$ will
transform the D.E. into $d^2y/dx^2 + y = 0$, since $\alpha = 1$ and
$\beta = 1$. Thus $y = c_1\cos x + c_2\sin x$, with $x = \ln t$, $t > 0$.

Section 3.5, Page 159

1. Substituting $y = e^{rt}$ into the D.E., we find that
$r^2 - 2r + 1 = 0$, which gives $r_1 = 1$ and $r_2 = 1$. Since the
roots are equal, the second linearly independent solution
is te^t and thus the general solution is $y = c_1e^t + c_2te^t$.

9. The characteristic equation is $25r^2 - 20r + 4 = 0$, which
may be written as $(5r-2)^2 = 0$ and hence the roots are
$r_1, r_2 = 2/5$. Thus $y = c_1e^{2t/5} + c_2\, te^{2t/5}$.

14. The characteristic equation is
$r^2 + 4r + 4 = (r+2)^2 = 0$, which
has the repeated root $r = -2$.
Since the I.C. are given at
$t = -1$, write the general solution
as $y = c_1e^{-2(t+1)} + c_2te^{-2(t+1)}$. Then

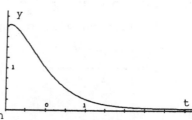

$$y' = -2c_1e^{-2(t+1)} + c_2e^{-2(t+1)} - 2c_2te^{-2(t+1)} \text{ and hence}$$

$c_1 - c_2 = 2$ and $-2c_1 + 3c_2 = 1$ which yield $c_1 = 7$ and $c_2 = 5$.
Thus $y = 7e^{-2(t+1)} + 5te^{-2(t+1)}$, a decaying exponential as shown in the graph.

17a. The characteristic equation is $4r^2 + 4r + 1 = (2r+1)^2 = 0$, so we have $y(t) = (c_1 + c_2t)e^{-t/2}$. Thus $y(0) = c_1 = 1$ and $y'(0) = -c_1/2 + c_2 = 2$ and hence $c_1 = 1$, $c_2 = 5/2$ and

$$y(t) = (1 + 5t/2)e^{-t/2}.$$

17b. From part (a), $y'(t) = -\dfrac{1}{2}(1 + 5t/2)e^{-t/2} + \dfrac{5}{2}e^{-t/2} = 0$, when

$-\dfrac{1}{2} - \dfrac{5t}{4} + \dfrac{5}{2} = 0$, or $t_0 = \dfrac{8}{5}$ and $y_0 = 5e^{-4/5}$.

17c. From part (a), $-\dfrac{1}{2} + c_2 = b$ or $c_2 = b + \dfrac{1}{2}$ and

$$y(t) = [1 + (b+\dfrac{1}{2})t]e^{-t/2}.$$

17d. From part (c), $y'(t) = -\dfrac{1}{2}[1 + (b+\dfrac{1}{2})t]e^{-t/2} + (b+\dfrac{1}{2})e^{-t/2} = 0$

which yields $t_M = \dfrac{4b}{2b+1}$ and

$$y_M = (1 + \dfrac{2b+1}{2} \cdot \dfrac{4b}{2b+1})e^{-2b/(2b+1)} = (1 + 2b)e^{-2b/(2b+1)}.$$

19. If $r_1 = r_2$ then $y(t) = (c_1 + c_2t)e^{r_1t}$. Since the exponential is never zero, $y(t)$ can be zero only if $c_1 + c_2t = 0$, which yields at most one value of t if c_1 and c_2 differ in sign. If $r_1 \neq r_2$, then

$$y(t) = c_1e^{r_1t} + c_2e^{r_2t} = e^{r_1t}(c_1 + c_2e^{(r_2-r_1)t}).$$ Again, this is zero only if c_1 and c_2 differ in sign, in which case

$$t = \dfrac{\ln(-c_1/c_2)}{(r_2-r_1)}.$$

21. If $r_2 \neq r_1$ then $\phi(t; r_1, r_2) = (e^{r_2 t} - e^{r_1 t})/(r_2 - r_1)$ is
    defined for all t. Note that ϕ is a linear combination
    of two solutions, $e^{r_1 t}$ and $e^{r_2 t}$, of the D.E. Hence, ϕ is a
    solution of the differential equation. Think of r_1 as
    fixed and let $r_2 \to r_1$. The limit of ϕ as $r_2 \to r_1$ is
    indeterminate. If we use L'Hopital's rule, we find

    $$\lim_{r_2 \to r_1} \frac{e^{r_2 t} - e^{r_1 t}}{r_2 - r_1} = \lim_{r_2 \to r_1} \frac{te^{r_2 t}}{1} = te^{r_1 t}. \quad \text{Hence, the}$$

    solution $\phi(t; r_1, r_2) \to te^{r_1 t}$ as $r_2 \to r_1$.

25. Let $y_2 = v/t$. Then $y_2' = v'/t - v/t^2$ and
    $y_2'' = v''/t - 2v'/t^2 + 2v/t^3$. Substituting in the D.E. we
    obtain
    $t^2(v''/t - 2v'/t^2 + 2v/t^3) + 3t(v'/t - v/t^2) + v/t = 0$.
    Simplifying the left side we get $tv'' + v' = 0$, which
    yields $v' = c_1/t$. Thus $v = c_1 \ln t + c_2$. Hence a second
    solution is $y_2(t) = (c_1 \ln t + c_2)/t$. However, we may set
    $c_2 = 0$ and $c_1 = 1$ without loss of generality and thus we
    have $y_2(t) = (\ln t)/t$ as a second solution. Note that in
    the form we actually calculated, $y_2(t)$ is a linear
    combination of $1/t$ and $\ln t/t$, and hence is the general
    solution.

27. In this case the calculations are somewhat easier if we
    do not use the explicit form for $y_1(x) = \sin x^2$ at the
    beginning but simply set $y_2(x) = y_1 v$. Substituting this
    form for y_2 in the D.E. gives $x(y_1 v)'' - (y_1 v)' + 4x^3(y_1 v) = 0$.
    On carrying out the differentiations and making use of
    the fact that y_1 is a solution, we obtain
    $xy_1 v'' + (2xy_1' - y_1)v' = 0$. This is a first order linear
    equation for v', which has the solution $v' = cx/(\sin x^2)^2$.
    Setting $u = x^2$ allows integration of this to get
    $v = c_1 \cot x^2 + c_2$. Setting $c_1 = 1$, $c_2 = 0$ and multiplying
    by $y_1 = \sin x^2$ we obtain $y_2(x) = \cos x^2$ as the second
    solution of the D.E.

30. Substituting $y_2(x) = y_1(x)v(x)$ in the D.E. gives
$x^2(y_1v)'' + x(y_1v)' + (x^2 - \frac{1}{4})y_1v = 0$. On carrying out the
differentiations and making use of the fact that y_1 is a
solution, we obtain $x^2y_1v'' + (2x^2y_1' + xy_1)v' = 0$. This is
a first order linear equation for v',
$v'' + (2y_1'/y_1 + 1/x)v' = 0$, with solution

$$v'(x) = c\exp[-\int(2\frac{y_1'}{y_1} + \frac{1}{x})dx] = c\exp[-2\ln y_1 - \ln x]$$

$$= c\frac{1}{xy_1{}^2} = \frac{c}{x(x^{-1}\sin^2 x)} = c\csc^2 x,$$

where c is an arbitrary constant, which we will take to
be one. Then $v(x) = \int\csc^2 x\, dx = -\cot x + k$ where again k
is an arbitrary constant which can be taken equal to
zero. Thus
$y_2(x) = y_1(x)v(x) = (x^{-1/2}\sin x)(-\cot x) = -x^{-1/2}\cos x$. The
second solution is usually taken to be $x^{-1/2}\cos x$. Note
that c = -1 would have given this solution.

31b. Let $y_2(x) = e^x v(x)$, then $y_2' = e^x v' + e^x v$, and
$y_2'' = e^x v'' + 2e^x v' + e^x v$. Substituting in the D.E. we
obtain $xe^x v'' + (xe^x - Ne^x)v' = 0$, or $v'' + (1-N/x)v' = 0$.
This is a first order linear D.E. for v' with integrating
factor $\mu(x) = \exp[\int(1-N/x)dx] = x^{-N}e^x$. Hence
$(x^{-N}e^x v')' = 0$, and $v' = cx^N e^{-x}$ which gives
$v(x) = c\int x^N e^{-x}dx + k$. On taking k = 0 we obtain as the
second solution $y_2(x) = ce^x\int x^N e^{-x}dx$. The integral can be
evaluated by using the method of integration by parts.
At each stage let $u = x^N$ or x^{N-1}, or whatever the power of
x that remains, and let $dv = e^{-x}$. Note that this dv is
not related to the $v(x)$ in $y_2(x)$. For N = 2 we have

$$y_2(x) = ce^x\int x^2 e^{-x}dx = ce^x[x^2\frac{e^{-x}}{-1} - \int 2x\frac{e^{-x}}{-1}dx]$$

$$= -cx^2 + ce^x[2x\frac{e^{-x}}{-1} - \int 2\frac{e^{-x}}{-1}dx]$$

$$= c(-x^2 - 2x - 2) = -2c(1 + x + x^2/2!).$$

Choosing c = -1/2! gives the desired result. For the
general case c = - 1/N!

33. $(y_2/y_1)' = (y_1 y_2' - y_1' y_2)/y_1^2 = W(y_1, y_2)/y_1^2$. Abel's identity

is $W(y_1, y_2) = c\ \exp[-\int_{t_0}^{t} p(r)\,dr]$. Hence

$(y_2/y_1)' = c y_1^{-2} \exp[-\int_{t_0}^{t} p(r)\,dr]$. Integrating and setting

$c = 1$ (since a solution y_2 can be multiplied by any
constant) and taking the constant of integration to be
zero we obtain

$$y_2(t) = y_1(t) \int_{t_0}^{t} \frac{\exp[-\int_{s_0}^{s} p(r)\,dr]}{[y_1(s)]^2}\ ds.$$

35. From Problem 33 and Abel's formula we have

$$(\frac{y_2}{y_1})' = \frac{\exp[\int (1/t)\,dt]}{\sin^2(t^2)} \cdot\ = \frac{e^{\ln t}}{\sin^2(t^2)} \cdot\ = t\csc^2(t^2).$$ Thus

$y_2/y_1 = -(1/2)\cot(t^2)$ and hence we can choose $y_2 = \cos(t^2)$

since $y_1 = \sin^2(t^2)$.

38. The general solution of the D.E. is $y = c_1 e^{r_1 t} + c_2 e^{r_2 t}$

where $r_1, r_2 = (-b \pm \sqrt{b^2 - 4ac})/2a$ provided $b^2 - 4ac \neq 0$.

In this case there are two possibilities. If $b^2 - 4ac > 0$

then $(b^2 - 4ac)^{1/2} < b$ and r_1 and r_2 are real and

negative. Consequently $e^{r_1 t} \to 0$ and $e^{r_2 t} \to 0$; and hence

$y \to 0$, as $t \to \infty$. If $b^2 - 4ac < 0$ then r_1 and r_2 are

complex conjugates with negative real part. Again

$e^{r_1 t} \to 0$ and $e^{r_2 t} \to 0$; and hence $y \to 0$, as $t \to \infty$.

Finally, if $b^2 - 4ac = 0$, then $y = c_1 e^{r_1 t} + c_2 t e^{r_1 t}$ where

$r_1 = -b/2a < 0$. Hence, again $y \to 0$ as $t \to \infty$. This

conclusion does not hold if either $b = 0$ or $c = 0$.

42. Substituting $z = \ln t$ into the D.E. gives

$\dfrac{d^2 y}{dz^2} + \dfrac{dy}{dz} + 0.25y = 0$, which has the solution

$y(z) = c_1 e^{-z/2} + c_2 z e^{-z/2}$ so that $y(t) = c_1 t^{-1/2} + c_2 t^{-1/2} \ln t$.

Section 3.6, Page 171

1. First we find the solution of the homogeneous D.E., which
    has the characteristic equation $r^2-2r-3 = (r-3)(r+1) = 0$.
    Hence $y_c = c_1 e^{3t} + c_2 e^{-t}$ and we can assume $Y = Ae^{2t}$ for the
    particular solution. Thus $Y' = 2Ae^{2t}$ and $Y'' = 4Ae^{2t}$ and
    substituting into the D.E. yields
    $4Ae^{2t} - 2(2Ae^{2t}) - 3(Ae^{2t}) = 3e^{2t}$. Thus $-3A = 3$ and
    $A = -1$, yielding $y = c_1 e^{3t} + c_2 e^{-t} - e^{2t}$.

4. Initially we assume $Y = A + B\sin 2t + C\cos 2t$. However,
    since a constant is a solution of the related homogeneous
    D.E. we must modify Y by multiplying the constant A by t
    and thus the correct form is $Y = At + B\sin 2t + C\cos 2t$.

6. Since $y_c = c_1 e^{-t} + c_2 t e^{-t}$ we must assume $Y = At^2 e^{-t}$, so
    that $Y' = 2Ate^{-t} - At^2 e^{-t}$ and $y'' = 2Ae^{-t} - 4Ate^{-t} + At^2 e^{-t}$.
    Substituting in the D.E. gives $(At^2-4At+2A)e^{-t} +$
    $2(-At^2+2At)e^{-t} + At^2 e^{-t} = 2e^{-t}$. Notice that all terms on
    the left involving t^2 and t add to zero and we are left
    with $2A = 2$, or $A = 1$. Hence $y = c_1 e^{-t} + c_2 t e^{-t} + t^2 e^{-t}$.

8. The assumed form is $Y = (At + B)\sin 2t + (Ct + D)\cos 2t$,
    which is appropriate for both terms appearing on the
    right side of the D.E. Since none of the terms appearing
    in Y are solutions of the homogeneous equation, we do not
    need to modify Y.

11. First solve the homogeneous D.E. Substituting $y = e^{rt}$
    gives $r^2 + r + 4 = 0$. Hence $y_c = e^{-t/2}[c_1\cos(\sqrt{15}\,t/2) +$
    $c_2\sin(\sqrt{15}\,t/2)]$. We replace $\sinh t$ by $(e^t - e^{-t})/2$ and
    then assume $Y(t) = Ae^t + Be^{-t}$. Since neither e^t nor e^{-t}
    are solutions of the homogeneous equation, there is no
    need to modify our assumption for Y. Substituting in the
    D.E., we obtain $6Ae^t + 4Be^{-t} = e^t - e^{-t}$. Hence, $A = 1/6$
    and $B = -1/4$. The general solution is
    $y = e^{-t/2}[c_1\cos(\sqrt{15}\,t/2) + c_2\sin(\sqrt{15}\,t/2)] + e^t/6 - e^{-t}/4$.
    [For this problem we could also have found a particular

solution as a linear combination of sinht and cosht:
$Y(t)$ = Acosht + Bsinht. Substituting this in the D.E.
gives
(5A + B)cosht + (A + 5B)sinht = 2sinht. The solution is
A = -1/12 and B = 5/12. A simple calculation shows that
$-(1/12)$cosht + (5/12)sinht = $e^t/6 - e^{-t}/4$.]

13. $y_c = c_1e^{-2t} + c_2e^t$ so for the particular solution we
assume Y = At + B. Since neither At or B are solutions of
the homogeneous equation it is not necessary to modify
the original assumption. Substituting Y in the D.E. we
obtain 0 + A -2(At+B) = 2t or -2A = 2 and A-2B = 0.
Solving for A and B we obtain $y = c_1e^{-2t} + c_2e^t - t - 1/2$
as the general solution. $y(0) = 0 \rightarrow c_1 + c_2 - 1/2 = 0$
and $y'(0) = 1 \rightarrow -2c_1 + c_2 - 1 = 1$, which yield $c_1 = -1/2$
and $c_2 = 1$. Thus $y = e^t - (1/2)e^{-2t} - t - 1/2$.

16. Since the nonhomogeneous term is the product of a linear
polynomial and an exponential, assume Y of the same form:
$Y = (At+B)e^{2t}$.

19a. The solution of the homogeneous D.E. is $y_c = c_1e^{-3t} + c_2$.
After inspection of the nonhomogeneous term, we assume
$Y(t) = (A_0t^4 +A_1t^3 +A_2t^2 +A_3t+A_4) + (B_0t^2 +B_1t+B_2)e^{-3t} + C$sin3t +
Dcos3t. However, since e^{-3t} and a constant are solutions
of the homogeneous D.E., we must multiply the coefficient
of e^{-3t} and the polynomial by t. The correct form is
$Y(t) = t(A_0t^4 + A_1t^3 + A_2t^2 + A_3t + A_4) +$
$t(B_0t^2 + B_1t + B_2)e^{-3t} + C$sin3t + Dcos3t.

22a. The solution of the homgeneous D.E. is $y_c = e^{-t}[c_1$cost +
$c_2$sint]. After inspection of the nonhomogeneous term, we
assume $Y(t) = Ae^{-t} + (B_0t^2 + B_1t + B_2)e^{-t}$cost + $(C_0t^2 + C_1t$
+ $C_2)e^{-t}$sint. Since e^{-t}cost and e^{-t}sint are solutions of
the homogeneous D.E., it is necessary to multiply both
the last two terms by t. Hence the correct form is
$Y(t) = Ae^{-t} + t(B_0t^2 + B_1t + B_2)e^{-t}$cost +
$t(C_0t^2 + C_1t + C_2)e^{-t}$sint.

28. First solve the I.V.P. $y'' + y = t$, $y(0) = 0$, $y'(0) = 1$
    for $0 \leq t \leq \pi$. The solution of the homogeneous D.E. is
    $y_c(t) = c_1\cos t + c_s\sin t$. The correct form for $Y(t)$ is
    $y(t) = A_0 t + A_1$. Substituting in the D.E. we find $A_0 = 1$
    and $A_1 = 0$. Hence, $y = c_1\cos t + c_2\sin t + t$. Applying the
    I.C., we obtain $y = t$. For $t > \pi$ the form for $Y(t)$ is
    $Y(t) = Ee^{\pi-t}$. Substituting $Y(t)$ in the D.E., we obtain
    $Ee^{\pi-t} + Ee^{\pi-t} = \pi e^{\pi-t}$ so $E = \pi/2$. Hence the general
    solution for $t > \pi$ is $Y = D_1\cos t + D_2\sin t + (\pi/2)e^{\pi-t}$. If
    y and y' are to be continuous at $t = \pi$, then the
    solutions and their derivatives for $t \leq \pi$ and $t > \pi$ must
    have the same value at $t = \pi$. These conditions require
    $\pi = -D_1 + \pi/2$ and $1 = -D_2 - \pi/2$. Hence $D_1 = -\pi/2$,
    $D_2 = -(1 + \pi/2)$, and

$$y = \phi(t) = \begin{cases} t, & 0 \leq t \leq \pi \\ -(\pi/2)\cos t - (1 + \pi/2)\sin t + (\pi/2)e^{\pi-t}, & t > \pi. \end{cases}$$

The graphs of the nonhomogeneous term and ϕ follow.

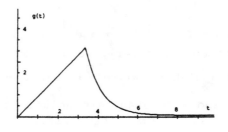 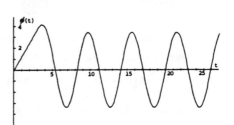

30. According to Theorem 3.6.1, the difference of any two
    solutions of the linear second order nonhomogeneous D.E.
    is a solution of the corresponding homogeneous D.E.
    Hence $Y_1 - Y_2$ is a solution of $ay'' + by' + cy = 0$. In
    Problem 38 of Section 3.5 we showed that if $a > 0$, $b > 0$,
    and $c > 0$ then every solution of this D.E. goes to zero
    as $t \to \infty$.

33. From Problem 32 we write the D.E. as $(D-4)(D+1)y = 3e^{2t}$.
    Thus let $(D+1)y = u$ and then $(D-4)u = 3e^{2t}$. This last
    equation is the same as $du/dt - 4u = 3e^{2t}$, which may be
    solved by multiplying both sides by e^{-4t} and integrating

(see section 2.1). This yields $u = (-3/2)e^{2t} + Ce^{4t}$.
Substituting this form of u into $(D+1)y = u$ we obtain
$dy/dt + y = (-3/2)e^{2t} + Ce^{4t}$. Again, multiplying by e^{t}
and integrating gives $y = (-1/2)e^{2t} + C_1e^{4t} + C_2e^{-t}$, where
$C_1 = C/5$.

Section 3.7, Page 177

2. Two linearly independent solutions of the homogeneous
    D.E. are $y_1(t) = e^{2t}$ and $y_2(t) = e^{-t}$. Assume $Y = u_1(t)e^{2t}$
    $+ u_2(t)e^{-t}$, then $Y'(t) = [2u_1(t)e^{2t} - u_2(t)e^{-t}] + [u_1'(t)e^{2t}$
    $+ u_2'(t)e^{-t}]$. We set $u_1'(t)e^{2t} + u_2'(t)e^{-t} = 0$. Computing
    $Y''(t)$ and substituting in the D.E. gives
    $2u_1'(t)e^{2t} - u_2'(t)e^{-t} = 2e^{-t}$. Thus we have two algebraic
    equations for $u_1'(t)$ and $u_2'(t)$ with the solution
    $u_1'(t) = 2e^{-3t}/3$ and $u_2'(t) = -2/3$. Hence $u_1(t) = -2e^{-3t}/9$
    and $u_2(t) = -2t/3$. Substituting in the formula for $Y(t)$
    we obtain $Y(t) = (-2e^{-3t}/9)e^{2t} + (-2t/3)e^{-t} = (-2e^{-t}/9) -$
    $(2te^{-t}/3)$. Since e^{-t} is a solution of the homogeneous
    D.E., we can choose $Y(t) = -2te^{-t}/3$.

5. Since cost and sint are solutions of the homogeneous
    D.E., we assume $Y = u_1(t)\cos t + u_2(t)\sin t$. Thus
    $y' = -u_1(t)\sin t + u_2(t)\cos t$ after setting
    $u_1'(t)\cos t + u_2'(t)\sin t = 0$. Finding Y'' and substituting
    into the D.E. then yields $-u_1'(t)\sin t + u_2'(t)\cos t = \tan t$.
    The two equations for $u_1'(t)$ and $u_2'(t)$ have the
    solution: $u_1'(t) = -\sin^2 t/\cos t = -\sec t + \cos t$ and
    $u_2'(t) = \sin t$. Thus $u_1(t) = \sin t - \ln(\tan t + \sec t)$ and
    $u_2(t) = -\cos t$, which when substituted into the assumed
    form for y, simplified, and added to the homogeneous
    solution yields
    $y = c_1\cos t + c_2\sin t - (\cos t)\ln(\tan t + \sec t)$.

11. Two linearly independent solutions of the homogeneous
    D.E. are $y_1(t) = e^{3t}$ and $y_2(t) = e^{2t}$. Applying Theorem

3.7.1 with $W(y_1, y_2)(t) = -e^{5t}$, we obtain

$$Y(t) = -e^{3t}\int \frac{e^{2s}g(s)}{-e^{5s}}\,ds + e^{2t}\int \frac{e^{3s}g(s)}{-e^{5s}}\,ds$$

$$= \int [e^{3(t-s)} - e^{2(t-s)}]g(s)\,ds.$$

The complete solution is then obtained by adding $c_1 e^{3t} + c_2 e^{2t}$ to $Y(t)$.

14. That t and te^t are solutions of the homogeneous D.E. can be verified by direction substitution. Thus we assume $Y = xu_1(t) + te^t u_2(t)$. Following the pattern of earlier problems we find $t\,u_1'(t) + te^t u_2'(t) = 0$ and $u_1'(t) + (t+1)e^t u_2' = 2t$. [Note that $g(t) = 2t$, since the D.E. must be put into the form of Eq.(16)]. The solution of these equations gives $u_1'(t) = -2$ and $u_2'(t) = 2e^{-t}$. Hence, $u_1(t) = -2t$ and $u_2(t) = -2e^{-t}$, and $Y(t) = t(-2t) + te^t(-2e^{-t}) = -2t^2 - 2t$. However, since t is a solution of the homogeneous D.E. we can choose as our particular solution $Y(t) = -2t^2$.

18. For this problem, and for many others, it is probably easier to rederive Eqs.(26) without using the explicit form for $y_1(x)$ and $y_2(x)$ and then to substitute for $y_1(x)$ and $y_2(x)$ in Eqs.(26). In this case if we take $y_1 = x^{-1/2}\sin x$ and $y_2 = x^{-1/2}\cos x$, then $W(y_1, y_2) = -1/x$. If the D.E. is put in the form of Eq.(16), then $g(x) = 3x^{-1/2}\sin x$ and thus $u_1'(x) = 3\sin x\cos x$ and $u_2'(x) = -3\sin^2 x = 3(-1 + \cos 2x)/2$. Hence $u_1(x) = (3\sin^2 x)/2$ and $u_2(x) = -3x/2 + 3(\sin 2x)/4$, and

$$Y(x) = \frac{3\sin^2 x}{2}\frac{\sin x}{\sqrt{x}} + \left(-\frac{3x}{2} + \frac{3\sin 2x}{4}\right)\frac{\cos x}{\sqrt{x}}$$

$$= \frac{3\sin^2 x}{2}\frac{\sin x}{\sqrt{x}} + \left(-\frac{3x}{2} + \frac{3\sin x\cos x}{2}\right)\frac{\cos x}{\sqrt{x}}$$

$$= \frac{3\sin x}{2\sqrt{x}} - \frac{3\sqrt{x}\cos x}{2}.$$

The first term is a multiple of $y_1(x)$ and thus can be neglected for $Y(x)$.

22. Putting limits on the integrals of Eq.(28) and changing the integration variable to s yields

$$Y(t) = -y_1(t) \int_{t_0}^t \frac{y_2(s)g(s)ds}{W(y_1,y_2)(s)} + y_2(t) \int_{t_0}^t \frac{y_1(s)g(s)ds}{W(y_1,y_2)(s)}$$

$$= \int_{t_0}^t \frac{-y_1(t)y_2(s)g(s)ds}{W(y_1,y_2)(s)} + \int_{t_0}^t \frac{y_2(t)y_1(s)g(s)ds}{W(y_1,y_2)(s)}$$

$$= \int_{t_0}^t \frac{[y_1(s)y_2(t) - y_1(t)y_2(s)]g(s)ds}{y_1(s)y_2'(s) - y_1'(s)y_2(s)}.$$

25. Note that $y_1 = e^{\lambda t}\cos\mu t$ and $y_2 = e^{\lambda t}\sin\mu t$ and thus $W(y_1,y_2) = \mu e^{2\lambda t}$. From Problem 22 we then have:

$$Y(t) = \int_{t_0}^t \frac{e^{\lambda s}\cos\mu s e^{\lambda t}\sin\mu t - e^{\lambda t}\cos\mu t e^{\lambda s}\sin\mu s}{\mu e^{2\lambda s}} g(s)ds$$

$$= \mu^{-1} \int_{t_0}^t e^{\lambda(t-s)}[\cos\mu s \sin\mu t - \cos\mu t \sin\mu s]g(s)ds$$

$$= \mu^{-1} \int_{t_0}^t e^{\lambda(t-s)}[\sin\mu(xt-s)]g(s)ds.$$

29. First, we put the D.E. in standard form by dividing by t^2: $y'' - 2y'/t + 2y/t^2 = 4$. Assuming that $y = tv(t)$ and substituting in the D.E. we obtain $tv'' = 4$. Hence $v'(t) = 4\ln t + c_2$ and $v(t) = 4 \int \ln t \, dt + c_2 t = 4(t\ln t - t) + c_2 t$. The general solution is $c_1 y_1(t) + tv(t) = c_1 t + 4(t^2\ln t - t^2) + c_2 t^2$. Since $-4t^2$ is a multiple of $y_2 = c_2 t^2$ we can write $y = c_1 t + c_2 t^2 t^2 + 4t^2 \ln t$.

Section 3.8, Page 190

2. From Eq.(15) we have $R\cos\delta = -1$, and $R\sin\delta = \sqrt{3}$. Thus $R = \sqrt{1+3} = 2$ and $\delta = \tan^{-1}(-\sqrt{3}) + \pi = 2\pi/3 \cong 2.09440$. Note that we have to "add" π to the inverse tangent value since δ must be a second quadrant angle. Thus $u = 2\cos(t-2\pi/3)$.

6. The motion is an undamped free vibration. The units are in the CGS system. The spring constant $k = (100 \text{ gm})(980\text{cm/sec}^2)/5\text{cm}$. Hence the D.E. for the motion is $100u'' + [(100 \cdot 980)/5]u = 0$ where u is measured in cm and time in sec. We obtain $u'' + 196u = 0$ so

u = Acos14t + Bsin14t. The I.C. are u(0) = 0 → A = 0
and u'(0) = 10 cm/sec → B = 10/14 = 5/7. Hence
u(t) = 5/7sin4t, which first reaches equilibrium when
14t = π, or t = π/14.

8. We use Eq.(31) without R (there is no resistor) and E(t)
and with L = 1 henry and 1/C = 4×10^6 since C = $.25 \times 10^{-6}$
farads. Thus the I.V.P. is Q" + 4×10^6 Q = 0, Q(0) = 10^{-6}
coulombs and Q'(0) = 0.

9. The spring constant is k = (20)(980)/5 = 3920 dyne/cm.
The I.V.P. for the motion is 20u" + 400u' + 3920u = 0 or
u" + 20u' + 196u = 0 and u(0) = 2, u'(0) = 0. Here u is
measured in cm and t in sec. The general solution of the
D.E. is u = $Ae^{-10t}cos4\sqrt{6}\,t$ + $Be^{-10t}sin4\sqrt{6}\,t$. The I.C.
u(0) = 2 → A = 2 and u'(0) = 0 → -10A + $4\sqrt{6}$ B = 0. The
solution is u = e^{-10t}[2cos$4\sqrt{6}\,t$ + 5(sin$4\sqrt{6}\,t$)/$\sqrt{6}$]cm.
The quasi frequency is μ = $4\sqrt{2}$, the quasi period is
T_d = 2πμ = π/$2\sqrt{6}$ and T_d/T = 7/$2\sqrt{6}$ since
T = 2π/14 = π/7. To find an upper bound for τ, write u in
the form of Eq.(29): u(t) = $\sqrt{4+25/6}\,e^{-10t}$cos($4\sqrt{6}\,t\delta$). Now,
since |cos($4\sqrt{6}\,t-\delta$)| ≤ 1, we have |u(t)| < .05 ⇒
$\sqrt{4+25/6}\,e^{-10t}$ < .05, which yields τ = .4046. A more precise
answer can be obtained with a computer algebra system, which
in this case yields τ = .4045. The original estimate was
unusually close for this problem since cos($4\sqrt{6}\,t-\delta$) = -0.9996
for t = .4046.

12. Substituting the given values for L, C and R we obtain
the D.E. .2Q" + 3×10^2 Q' + 10^5 Q = 0. The I.C. are
Q(0) = 10^{-6} and Q'(0) = I(0) = 0. Assuming Q = e^{rt}, we
obtain the roots of the characteristic equation as
r_1 = -500 and r_2 = -1000. Thus Q = c_1e^{-500t} + c_2e^{-1000t}
and hence Q(0) = 10^{-6} → c_1 + c_2 = 10^{-6} and
Q'(0) = 0 → -500c_1 - 1000c_2 = 0. Solving for c_1 and c_2
yields the solution.

17. The mass is 8/32 lb-sec^2/ft, and the spring constant is
8/(1/8) = 64 lb/ft. Hence (1/4)u" + γu' + 64u = 0 or
u" + 4γu' + 256u = 0, where u is measured in ft, t in sec
and the units of γ are lb-sec/ft. We look for solutions
of the D.E. of the form u = e^{rt} and find
r^2 + 4γr + 256 = 0, so r_1,r_2 = [-4γ ± $\sqrt{16\gamma^2 - 1024}$]/2.
The system will be overdamped, critically damped or
underdamped as ($16\gamma^2$ - 1024) is > 0, =0, or < 0,

respectively. Thus the system is critically damped when
γ = 8 lb-sec/ft.

19. The general solution of the D.E. is $u = Ae^{r_1t} + Be^{r_2t}$
     where $r_1, r_2 = [-\gamma \pm (\gamma^2 - 4km)^{1/2}]/2m$ provided
     $\gamma^2 - 4km \neq 0$, and where A and B are determined by the
     I.C. When the motion is overdamped, $\gamma^2 - 4km > 0$ and
     $r_1 > r_2$. Setting u = 0, we obtain $Ae^{r_1t} = -Be^{r_2t}$ or
     $e^{(r_1-r_2)t} = -B/A$. Since the exponential function is a
     monotone function, there is at most one value of t
     (when B/A < 0) for which this equation can be satisfied.
     Hence u can vanish at most once. If the system is
     critically damped, the general solution is
     $u(t) = (A + Bt)e^{-\gamma t/2m}$. The exponential function is never
     zero; hence u can vanish only if A + Bt = 0. If B = 0
     then u never vanishes; if B $\neq$ 0 then u vanishes once at
     t = -A/B provided A/B < 0.

20. The general solution of Eq.(21) for the case of critical
     damping is $u = (A + Bt)e^{-\gamma t/2m}$. The I.C. $u(0) = u_0 \rightarrow$
     $A = u_0$ and $u'(0) = u_0' \rightarrow u_0' = A(-\gamma/2m) + B$. Hence
     $u = [u_0 + (u_0' + \gamma u_0/2m)t]e^{-\gamma t/2m}$. If $u_0' = 0$, then
     $u = u_0(1 + \gamma t/2m)e^{-\gamma t/2m}$, which is never zero since γ and
     m are positive. By L'Hopital's Rule $u \rightarrow 0$ as $t \rightarrow \infty$.
     Finally for $u_0 > 0$, we want the condition which will
     insure that u = 0 at least once. Since the exponential
     function is never zero we require
     $u_0 + (u_0' + \gamma u_0/2m)t = 0$ at a positive value of t. This
     requires that $u_0' + \gamma u_0/2m \neq 0$ and that
     $t = -u_0(u_0' + \gamma u_0/2m)^{-1} > 0$. We know that $u_0 > 0$ so we must
     have $u_0' + \gamma u_0/2m < 0$ or $u_0' < -\gamma u_0/2m$.

23. From Problem 21: $\Delta = \dfrac{2\pi\gamma}{\mu(2m)} = T_d\gamma/2m$. Substituting the

     known values we find $\gamma = \dfrac{(1/2)(3)}{.3} = 5$ lb sec/ft.

24. From Eq.13 $\omega_0^2 = \dfrac{2k}{3}$ so $P = 2\pi/\sqrt{2k/3} = \pi \Rightarrow k=6$. Thus

     $u(t) = c_1\cos 2t + c_2\sin 2t$ and $u(0) = 2 \Rightarrow c_1 = 2$ and

$u'(0) = v \Rightarrow c_2 = v/2$. Hence $u(t) = 2\cos 2t + \dfrac{v}{2}\sin 2t =$

$\sqrt{4+\dfrac{v^2}{4}}\cos(2t-\gamma)$. Thus $\sqrt{4+\dfrac{v^2}{4}} = 3$ and $v = \pm 2\sqrt{5}$.

27. First, consider the static case. Let $\Delta\ell$
     denote the length of the block below
     the surface of the water. The weight
     of the block, which is a downward force,

     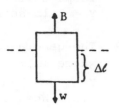

     is $w = \rho\ell^3 g$. This is balance by an equal
     and opposite buoyancy force B, which is
     equal to the weight of the displaced
     water. Thus $B = (\rho_0\ell^2\Delta\ell)g = \rho\ell^3 g$ so $\rho_0\Delta\ell = \rho\ell$. Now let x
     be the displacement of the block from its equilibrium
     position. We take downward as the positive direction.
     In a displaced position the forces acting on the blcok
     are its weight, which acts downward and is unchanged, and
     the buoyancy force which is now $\rho_0\ell^2(\Delta\ell + x)g$ and acts
     upward. The resultant force must be equal to the mass of
     the block times the acceleration, namely $\rho\ell^3 x''$. Hence
     $\rho\ell^3 g - \rho_0\ell^2(\Delta\ell + x)g = \rho\ell^3 x''$. The D.E. for the motion of
     the block is $\rho\ell^3 x'' + \rho_0\ell^2 gx = 0$. This gives a simple
     harmonic motion with frequency $(\rho_0 g/\rho\ell)^{1/2}$ and natural
     period $2\pi(\rho\ell/\rho_0 g)^{1/2}$.

29a. The characteristic equation is $4r^2 + r + 8 = 0$, so
     $r = (-1\pm\sqrt{127})/8$ and hence
     $$u(t) = e^{-t/8}(c_1\cos\frac{\sqrt{127}}{8}t + c_2\sin\frac{\sqrt{127}}{8} \quad c_2 = 2 \text{ or}$$
     $$c_2 = \frac{16}{\sqrt{127}}. \text{ Thus } u(t) = \frac{16}{\sqrt{127}}e^{-t/8}\sin\frac{\sqrt{127}}{8}t.$$

29c. The phase plot is the
     spiral shown and the
     direction of motion is
     clockwise since the
     graph starts at (0,2) and
     u increases initially.

     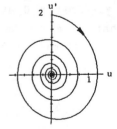

30c. Using u(t) as found in part (b), show that
     $ku^2/2 + m(u')^2/2 = (ka^2 + mb^2)/2$ for all t.

Section 3.9, Page 198

1.   We use the trigonometric identities
     $\cos(A \pm B) = \cos A \cos B \pm \sin A \sin B$ to obtain
     $\cos(A + B) - \cos(A - B) = -2\sin A \sin B$. If we choose
     $A + B = 9t$ and $A - B = 7t$, then $A = 8t$ and $B = t$.
     Substituting in the formula just derived, we have

$\cos 9t - \cos 7t = -2\sin 8t \sin t.$

5.   The mass $m = 4/32 = 1/8$ lb-sec^2/ft and the spring constant $k = 4/(1/8) = 32$ lb/ft. Since there is no damping, the I.V.P. is $(1/8)u'' + 32u = 2\cos 3t$, $u(0) = 1/6$, $u'(0) = 0$ where u is measured in ft and t in sec.

7a.   From the solution to Problem 5, we have $m = 1/8$, $F_0 = 2$, $\omega_0^2 = 256$, and $\omega^2 = 9$, so Eq.(3) becomes

$u = c_1\cos 16t + c_2\sin 16t + \dfrac{16}{247}\cos 3t.$ The I.C.

$u(0) = 1/6 \rightarrow c_1 + 16/247 = 1/6$ and $u'(0) = 0 \rightarrow 16c_2 = 0$, so the solution is $u = (151/1482)\cos 16t + (16/247)\cos 3t$ ft.

7c.   Resonance occurs when the frequency ω of the forcing function $4\sin\omega t$ is the same as the natural frequency ω_0 of the system. Since $\omega_0 = 16$, the system will resonate when $\omega = 16$.

10.   Note that this problem involves resonance and thus the particular solution has the form $t(A\cos 8t + B\sin 8t)$.

11a. For this problem the mass $m = 8/32$ lb-sec^2/ft and the spring constant $k = 8/(1/2) = 16$ lb/ft, so the D.E. is $0.25u'' + 0.25u' + 16u = 4\cos 2t$ where u is measured in ft and t in sec. To determine the steady state response we need only compute a particular solution of the nonhomogeneous D.E. since the solutions of the homogeneous D.E. decay to zero as $t \rightarrow \infty$. We assume $u(t) = A\cos 2t + B\sin 2t$, and substitute in the D.E.: $- A\cos 2t - B\sin 2t + (1/2)(-A\sin 2t + B\cos 2t) + 16(A\cos 2t + B\sin 2t) = 4\cos 2t$. Hence $15A + (1/2)B = 4$ and $- (1/2)A + 15B = 0$, from which we obtain $A = 240/901$ and $B = 8/901$. The steady state response is $u(t) = (240\cos 2t + 8\sin 2t)/901$.

11b. In order to determine the value of m that maximizes the steady state response, we note that the present problem has exactly the form of the problem considered in the text. Referring to Eqs.(8) and (9), the response is a maximum when $\Delta = f(m) = m^2(\omega_0^2 - \omega^2)^2 + \gamma^2\omega^2$, where $\omega_0^2 = k/m$, is a minimum. We calculate df/dm and set this quantity equal to zero to obtain $m = k/\omega^2$. We verify that this value of m gives a minimum of $f(m)$ by the second derivative test. For this problem $k = 16$ lb/ft and $\omega = 2$ rad/sec so the value of m that maximizes the response of the system is $m = 4$ slugs.

15. We must solve the three I.V.P.: $(1)\, u_1'' + u_1 = F_0 t/m,$ $0 < t < \pi,\ u_1(0) = u_1'(0) = 0;$ $(2)\, u_2'' + u_2 = F_0(2\pi-t)/m,$ $\pi < t < 2\pi,\ u_2(\pi) = u_1(\pi),\ u_2'(\pi) = u_1'(\pi);$ and $(3)\, u_3'' + u_3 = 0,\ 2\pi < t,\ u_3(2\pi) = u_2(2\pi),\ u_3'(2\pi) = u_2'(2\pi).$ The conditions at π and 2π insure the continuity of u and u' at those points. The general solutions of the D.E. are $u_1 = b_1\cos t + b_2\sin t + F_0 t/m,$ $u_2 = c_1\cos t + c_2\sin t + F_0(2\pi-t)/m,$ and $u_3 = d_1\cos t + d_2\sin t.$ The I.C. matching conditions, in order, give $b_1 = 0,\ b_2 + F_0/m = 0,$ $-b_1 + \pi F_0/m = -c_1 + \pi F_0/m,\ -b_2 + F_0/m = -c_2 - F_0/m,\ c_1 = d_1,$ and $c_2 - F_0/m = d_2.$ Solving these equations we obtain

$$u = (F_0/m) \begin{cases} t - \sin t & ,\ 0 \le t \le \pi \\ (2\pi - t) - 3\sin t & ,\ \pi < t \le 2\pi \\ -4\sin t & ,\ 2\pi < t. \end{cases}$$

16. The I.V.P. is $Q'' + 5\times10^3\, Q' + 4\times10^6\, Q = 12,\ Q(0) = 0,$ and $Q'(0) = 0.$ The particular solution is of the form $Q = A,$ so that upon substitution into the D.E. we obtain $4\times10^6 A = 12$ or $A = 3\times10^{-6}.$ The general solution of the D.E. is $Q = c_1 e^{r_1 t} + c_2 e^{r_2 t} + 3\times10^{-6},$ where r_1 and r_2 satisfy $r^2 + 5\times10^3 r + 4\times10^6 = 0$ and thus are $r_1 = -1000$ and $r_2 = -4000.$ The I.C. yield $c_1 = -4\times10^{-6}$ and $c_2 = 10^{-6}$ and thus $Q = 10^{-6}(e^{-4000t} - 4e^{-1000t} + 3)$ coulombs. Substituting $t = .001$ sec we obtain $Q(.001) = 10^{-6}(e^{-4} - 4e^{-1} + 3) = 1.5468 \times 10^{-6}$ coulombs. Since the exponentials are to a negative power $Q(t) \to 3\times10^{-6}$ coulombs as $t \to \infty$, which is the steady state charge.

22. The amplitude of the steady state response is seven or eight times the amplitude (3) of the forcing term. This large an increase is due to the fact that the forcing function has the same frequency as the natural frequency of the system.

There also appears to be a phase lag of approximately 1/4
of a period. That is, the maximum of the response occurs
1/4 of a period after the maximum of the forcing
function. Both these results are substantially different
than those of either Problems 21 or 23.

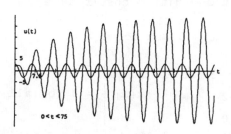
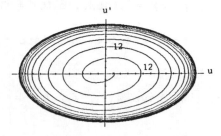

24.

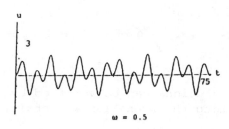
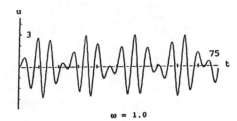

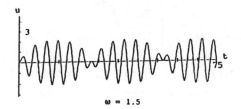
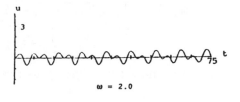

From viewing the above graphs, it appears that the system
exhibits a beat near $\omega = 1.5$, while the pattern for $\omega = 1.0$
is more irregular. However, the system exhibits the resonance
characteristic of the linear system for ω near 1, as the
amplitude of the response is the largest here.

CHAPTER 4

Section 4.1, Page 206

2. Writing the equation in standard form, we obtain
    $y''' + (sinx/t)y'' + (3/t)y = cost/t$. The functions
    $p_1(t) = sint/t$, $p_3(t) = 3/t$ and $g(t) = cost/t$ have
    discontinuities at $t = 0$. Hence Theorem 4.1.1 guarantees
    that a solution exists for $t < 0$ and for $t > 0$.

8. We have $w(f_1, f_2, f_3) = \begin{vmatrix} 2t-3 & 2t^2+1 & 3t^2+t \\ 2 & 4t & 6t+1 \\ 0 & 4 & 6 \end{vmatrix} = 0$ for all t.

    Thus by the extension of Theorem 3.31 the given functions
    are linearly dependent. Thus
    $c_1(2t-3) + c_2(2t^2+1) + c_3(3t^2+t) =$
    $(2c_2+3c_3)t^2 + (2c_1+c_3)t + (-3c_1+c_2) = 0$ when $c_1 = 1$ $c_2 = 3$
    and $c_3 = -2$.

13. That e^t, e^{-t}, and e^{-2t} are solutions can be verified by
    direct substitution. Computing the Wronskian we obtain,

    $$W(e^t, e^{-t}, e^{-2t}) = \begin{vmatrix} e^t & e^{-t} & e^{-2t} \\ e^t & -e^{-t} & -2e^{-2t} \\ e^t & e^{-t} & 4e^{-2t} \end{vmatrix} = -6e^{-2t}.$$

17. To show that the given Wronskian is zero, it is helpful
    to note that $(sin^2t)' = 2sintcost = sin2t$. This result
    can be obtained directly since $sin^2t = (1 - cos2t)/2 =$
    $\dfrac{1}{10}(5) + (-1/2)cos2t$ and hence sin^2t is a linear
    combination of 5 and $cos2t$. Thus the functions are
    linearly dependent and their Wronskian is zero.

19c. If we let $L[y] = y^{iv} - 5y'' + 4y$ and if we use the result
    of Problem 19b, we have $L[e^{rt}] = (r^4 - 5r^2 + 4)e^{rt}$. Thus
    e^{rt} will be a solution of the D.E. provided
    $(r^2-4)(r^2-1) = 0$. Solving for r, we obtain the four
    solutions e^t, e^{-t}, e^{2t} and e^{-2t}. Since
    $W(e^t, e^{-t}, e^{2t}, e^{-2t}) \neq 0$, the four functions form a
    fundamental set of solutions.

21. Comparing this D.E. to that of Problem 20 we see that $p_1(t) = 2$ and thus from the results of Problem 20 we have
$$w = ce^{-\int 2dt} = ce^{-2t}.$$

27. As in Problem 26, let $y = v(t)e^t$. Differentiating three times and substituting into the D.E. yields
$(2-t)e^t v''' + (3-t)e^t v'' = 0$. Dividing by $(2-t)e^t$ and letting $w = v''$ we obtain the first order separable equation $w' = -\dfrac{t-3}{t-2}w = (-1 + \dfrac{1}{t-2})w$. Separating t and w, integrating, and then solving for w yields
$w = v'' = c_1(t-2)e^{-t}$. Integrating this twice then gives
$v = c_1 te^{-t} + c_2 t + c_3$ so that $y = ve^t = c_1 t + c_2 te^t + c_3 e^t$, which is the complete solution, since it contains the given $y_1(t)$ and three constants.

Section 4.2, Page 214

2. If $-1 + i\sqrt{3} = Re^{i\theta}$, then $R = [(-1)^2 + (\sqrt{3})^2]^{1/2} = 2$. The angle θ is given by $R\cos\theta = 2\cos\theta = -1$ and $R\sin\theta = 2\sin\theta = \sqrt{3}$. Hence $\cos\theta = -1/2$ and $\sin\theta = \sqrt{3}/2$ which has the solution $\theta = 2\pi/3$. The angle θ is only determined up to an additive integer multiple of $\pm 2\pi$.

8. Writing $(1-i)$ in the form $Re^{i\theta}$, we obtain
$(1-i) = \sqrt{2}\, e^{i(-\pi/4+2m\pi)}$ where m is any integer. Hence,
$(1-i)^{1/2} = [2^{1/2}e^{i(-\pi/4+2m\pi)}]^{1/2} = 2^{1/4}e^{i(-\pi/8+m\pi)}$. We obtain the two square roots by setting $m = 0,1$. They are
$2^{1/4}e^{-i\pi/8}$ and $2^{1/4}e^{i7\pi/8}$. Note that any other integer value of m gives one of these two values. Also note that 1-i could be written as $1-i = \sqrt{2}\, e^{i(7\pi/4\, +\, 2m\pi)}$.

12. We look for solutions of the form $y = e^{rt}$. Substituting in the D.E., we obtain the characteristic equation
$r^3 - 3r^2 + 3r - 1 = 0$ which has roots $r = 1,1,1$. Since the roots are repeated, the general solution is
$y = c_1 e^t + c_2 te^t + c_3 t^2 e^t$.

15. We look for solutions of the form $y = e^{rt}$. Substituting in the D.E. we obtain the characteristic equation

$r^6 + 1 = 0$. The six roots of -1 are obtained by setting
$m = 0,1,2,3,4,5$ in $(-1)^{1/6} = e^{i(\pi+2m\pi)/6}$. They are
$e^{i\pi/6} = (\sqrt{3} + i)/2$, $e^{i\pi/2} = i$, $e^{i5\pi/6} = (-\sqrt{3} + i)/2$,
$e^{i7\pi/6} = (-\sqrt{3} - i)/2$, $e^{i3\pi/2} = -i$, and
$e^{i11\pi/6} = (\sqrt{3} - i)/2$. Note that there are three pairs of
conjugate roots. The general solution is
$$y = e^{\sqrt{3}t/2}[c_1\cos(t/2) + c_2\sin(t/2)] + e^{-\sqrt{3}t/2}[c_3\cos(t/2) + c_4\sin(t/2)] + c_5\cos t + c_6\sin t.$$

23. The characteristic equation is $r^3 - 5r^2 + 3r + 1 = 0$.
Using the procedure suggested following Eq. (12) we try
$r = 1$ as a root and find that indeed it is. Factoring
out $(r-1)$ we are then left with $r^2 - 4r - 1 = 0$, which
has the roots $2 \pm \sqrt{5}$.

27. The characteristic equation in this case is $12r^4 + 31r^3 + 75r^2 + 37r + 5 = 0$. Using an equation solver we find
$r = -\dfrac{1}{4}, -\dfrac{1}{3}, -1 \pm 2i$. Thus
$$y = c_1e^{-t/4} + c_2e^{-t/3} + e^{-t}(c_3\cos 2t + c_4\sin 2t).$$

29. The characteristic equation is $r^3 + r = 0$ and hence
$r = 0, +i, -i$ are the roots and the general solution is
$y(t) = c_1 + c_2\cos t + c_3\sin t$. $y(0) = 0$ implies
$c_1 + c_2 = 0$, $y'(0) = 1$ implies $c_3 = 1$ and $y''(0) = 2$
implies $-c_2 = 2$. Use this last equation in the first to
find $c_1 = 2$ and thus $y(t) = 2 - 2\cos t + \sin t$, which
continues to oscillate as $t \to \infty$.

30. The general solution is given by Eq. (21).

31. The general solution would normally be written
$y(t) = c_1 + c_2t + c_3e^{2t} + c_4te^{2t}$. However, in order to
evaluate the c's when the initial conditions are given at
$t = 1$, it is advantageous to rewrite
$y(t)$ as $y(t) = c_1 + c_2t + c_5e^{2(t-1)} + c_6(t-1)e^{2(t-1)}$.

34. The characteristic equation is $4r^3 + r + 5 = 0$, which has
roots $-1, \dfrac{1}{2} \pm i$. Thus

$$y(t) = c_1 e^{-t} + e^{-t/2}(c_2 \cos t + c_3 \sin t)$$

$$y'(t) = -c_1 e^{-t} + e^{-t/2}[(c_2/2 + c_3)\cos t + (-c_2 + c_3/2)\sin t]$$

and $y''(t) = c_1 e^{-t} + e^{-t/2}[(-3c_2/4 + c_3)\cos t +$
$(-c_2 - 3c_3/4)\sin t]$. The I.C. then yield
$c_1 + c_2 = 2$, $-c_1 + c_2/2 + c_3 = 1$ and $c_1 - 3c_2/4 + c_3 = -1$.
Solving these last three equations give $c_1 = 2/13$, $c_2 =$
$24/13$ and $c_3 = 3/13$.

37. The approach developed in this section for solving the
D.E. would normally yield $y(t) = c_1\cos t + c_2\sin t + c_5 e^t +$
$c_6 e^{-t}$ as the solution. Now use the definition of coshx
and sinht to yield the desired result. It is convenient
to use cosht and sinht rather than e^t and e^{-t} because the
I.C. are given at $t = 0$. Since cosht and sinht and all
of their derivatives are either 0 or 1 at $t = 0$, the
algebra in satisfying the I.C. is simplified.

38a. Since $p_1(t) = 0$, $W = ce^{-\int 0 dt} = c$.

39a. As in Section 3.8, the force that the spring designated
by k_1 exerts on mass m_1 is $-3u_1$. By an analysis similar
to that shown in Section 3.8, the middle spring exerts a
force of $-2(u_1-u_2)$ on mass m_1 and a force of $-2(u_2-u_1)$ on
mass m_2. In all cases the positive direction is taken in
the direction shown in Figure 4.2.4.

39c. From Eq.(i) we have $u_1''(0) = 2u_2(0) - 5u_1(0) = -1$ and
$u_1'''(0) = 2u_2'(0) - 5u_1'(0) = 0$. From Prob.39b we have
$u_1 = c_1\cos t + c_2\sin t + c_3\cos\sqrt{6}t + c_4\sin\sqrt{6}t$. Thus
$c_1+c_3 = 1$, $c_2+\sqrt{6}c_4 = 0$, $-c_1-6c_3 = -1$ and $-c_2-6\sqrt{6}c_4 = 0$,
which yield $c_1 = 1$ and $c_2 = c_3 = c_4 = 0$, so that
$u_1 = \cos t$. The first of Eqs.(i) then gives u_2.

Section 4.3, Page 219

1. First solve the homogeneous D.E. The characteristic
equation is $r^3 - r^2 - r + 1 = 0$, and the roots are $r = -1$,
1, 1; hence $y_c(t) = c_1 e^{-t} + c_2 e^t + c_3 t e^t$. Using the
superposition principle, we can write a particular

solution as the sum of particular solutions corresponding
to the D.E. $y''' - y'' - y' + y = 2e^{-t}$ and $y''' - y'' - y' + y = 3$. Our
initial choice for a particular solution, Y_1, of the
first equation is Ae^{-t}; but e^{-t} is a solution of the
homogeneous equation so we multiply by t. Thus,
$Y_1(t) = Ate^{-t}$. For the second equation we choose
$Y_2(t) = B$, and there is no need to modify this choice.
The constants are determined by substituting into the
individual equations. We obtain $A = 1/2$, $B = 3$. Thus,
the general solution is
$$y = c_1 e^{-t} + c_2 e^t + c_3 te^t + 3 + (te^{-t})/2.$$

9.   The characteristic equation for the related homogeneous
     D.E. is $r^3 + 4r = 0$ with roots $r = 0, +2i, -2i$. Hence
     $y_c(t) = c_1 + c_2 \cos 2t + c_3 \sin 2t$. The initial choice for
     $Y(t)$ is $At + B$, but since B is a solution of the
     homogeneous equation we must multiply by t and assume
     $Y(t) = t(At+B)$. A and B are found by substituting in the
     D.E., which gives $A = 1/8$, $B = 0$, and thus the general
     solution is $y(t) = c_1 + c_2 \cos 2t + c_3 \sin 2t + (1/8)t^2$.
     Applying the I.C. we have $y(0) = 0 \rightarrow c_1 + c_2 = 0$,
     $y'(0) = 0 \rightarrow 2c_3 = 0$, and $y''(0) = 1 \rightarrow -4c_2 + 1/4 = 1$,
     which have the solution $c_1 = 3/16$, $c_2 = -3/16$, $c_3 = 0$.
     For small t the graph will approximate $3(1 - \cos 2t)/16$ and
     for large t it will be approximated by $t^2/8$.

13. The characteristic equation for the homogeneous D.E. is
     $r^3 - 2r^2 + r = 0$ with roots $r = 0, 1, 1$. Hence the
     complementary solution is $y_c(t) = c_1 + c_2 e^t + c_3 te^t$. We
     consider the differential equations $y''' - 2y'' + y' = t^3$
     and $y''' - 2y'' + y' = 2e^t$ separately. Our initial choice
     for a particular solution, Y_1, of the first equation is
     $A_0 t^3 + A_1 t^2 + A_2 t + A_3$; but since a constant is a solution
     of the homogeneous equation we must multiply by t. Thus
     $Y_1(t) = t(A_0 t^3 + A_1 t^2 + A_2 t + A_3)$. For the second equation
     we first choose $Y_2(t) = Be^t$, but since both e^t and te^t
     are solutions of the homogeneous equation, we multiply by
     t^2 to obtain $Y_2(t) = Bt^2 e^t$. Then $Y(t) = Y_1(t) + Y_2(t)$ by
     the superposition principle and $y(t) = y_c(t) + Y(t)$.

17. The complementary solution is $y_c(t) = c_1 + c_2 e^{-t} + c_3 e^t + c_4 x e^t$. The superposition principle allows us to consider separately the D.E. $y^{iv} - y''' - y'' + y' = t^2 + 4$ and $y^{iv} - y''' - y'' + y' = t\sin t$. For the first equation our initial choice is $Y_1(t) = A_0 t^2 + A_1 t + A_2$; but this must be multiplied by t since a constant is a solution of the homogeneous D.E. Hence $Y_1(t) = t(A_0 t^2 + A_1 t + A_2)$. For the second equation our initial choice that $Y_2 = (B_0 t + B_1)\cos t + + (C_0 t + C_1)\sin t$ does not need to be modified. Hence $Y(t) = t(A_0 t^2 + A_1 t + A_2) + (B_0 t + B_1)\cos t + (C_0 t + C_1)\sin t$.

20. $(D-a)(D-b)f = (D-a)(Df-bf) = D^2 f - (a+b)Df + abf$ and $(D-b)(D-a)f = (D-b)(Df-af) = D^2 f - (b+a)Df + baf$. Since $a+b = b+a$ and $ab = ba$, we find the given equation holds for any function f.

22a. The D.E. of Problem 13 can be written as $D(D-1)^2 y = t^3 + 2e^t$. Since D^4 annihilates t^3 and $(D-1)$ annihilates $2e^t$, we have $D^5(D-1)^3 y = 0$, which corresponds to Eq.(ii) of Problem 21. The solution of this equation is $y(x) = A_1 t^4 + A_2 t^3 + A_3 t^2 + A_4 t + A_5 + (B_1 t^2 + B_2 t + B_3)e^{-t}$. Since $A_5 + (B_2 t + B_3)e^{-t}$ are solutions of the homogeneous equation related to the original D.E., they may be deleted and thus $Y(t) = A_1 t^4 + A_2 t^3 + A_3 t^2 + A_4 t + B_1 t^2 e^{-t}$.

22b. $(D+1)^2(D^2+1)$ annihilates the right side of the D.E. of Problem 14.

22e. $D^3(D^2+1)^2$ annihilates the right side of the D.E. of Problem 17.

Section 4.4, Page 224

1. The complementary solution is $y_c = c_1 + c_2\cos t + c_3\sin t$ and thus we assume a particular solution of the form $Y = u_1(t) + u_2(t)\cos t + u_3(t)\sin t$. Differentiating and assuming Eq.(5), we obtain $Y' = -u_2\sin t + u_3\cos t$ and
$$u_1' + u_2'\cos t + u_3'\sin t = 0 \qquad (a).$$

Continuing this process we obtain $Y'' = -u_2\cos t - u_3\sin t$, $Y''' = u_2\sin t - u_3\cos t - u_2'\cos t - u_3'\sin t$ and

$$-u_2'\sin t + u_3'\cos t = 0 \qquad (b).$$

Substituting Y and its derivatives, as given above, into the D.E. we obtain the third equation:

$$-u_2'\cos t - u_3'\sin t = \tan t \qquad (c).$$

Equations (a), (b) and (c) constitute Eqs.(10) of the text for this problem and may be solved to give $u_1' = \tan t$, $u_2' = -\sin t$, and $u_3' = -\sin^2 t/\cos t$. Thus $u_1 = -\ln\cos t$, $u_2 = \cos t$ and $u_3 = \sin t - \ln(\sec t + \tan t)$ and $Y = -\ln\cos t + 1 - (\sin t)\ln(\sec t + \tan t)$. Note that the constant 1 can be absorbed in c_1.

4. Replace tant in Eq. (c) of Prob. 1 by sect and use Eqs. (a) and (b) as in Prob. 1 to obtain $u_1' = \sec t$, $u_2' = -1$ and $u_3' = -\sin t/\cos t$.

5. Replace sect in Problem 7 with $e^{-t}\sin t$.

7. Since e^t, cost and sint are solutions of the related homogenous equation we have
    $Y(t) = u_1 e^t + u_2\cos t + u_3\sin t$. Eqs. (10) then are

    $$u_1'e^t + u_2'\cos t + u_3'\sin t = 0$$

    $$u_1'e^t - u_2'\sin t + u_3'\cos t = 0$$

    $$u_1'e^t - u_2'\cos t - u_3'\sin t = \sec t$$

    Since $W = 2e^t$, we have $u_1' = \dfrac{1}{2}e^{-2t}\sin t$

    $u_2' = -\dfrac{1}{2}e^{-t}\sin t(\cos t - \sin t)$ and $u_3' = -\dfrac{1}{2}e^{-t}\sin t(\sin t + \cos t)$.

    Using a computer algebra system to find u_1, u_2, and u_3, substitution into Y and simplifying yields

    $$Y = -\dfrac{1}{5}e^{-t}\cos t.$$

11. From Problem 7

    $y(0) = c_1 + c_2 = 2$, $y'(0) = c_1 + c_3 - \dfrac{1}{2} + \dfrac{1}{2} = -1$ and

    $y''(0) = c_1 - c_2 + \dfrac{1}{2} - 1 + \dfrac{1}{2} = 1$. Again, a computer

    algebra system may be used to yield the respective derivatives.

14. Since a fundamental set of solutions of the homogeneous D.E. is $y_1 = e^t$, $y_2 = \cos t$, $y_3 = \sin t$, a particular solution is of the form $Y(t) = e^t u_1(t) + (\cos t)u_2(t) + (\sin t)u_3(t)$. Differentiating and making the same assumptions that lead to Eqs.(10), we obtain

$$u_1' e^t + u_2' \cos t + u_3' \sin t = 0$$
$$u_1' e^t - u_2' \sin t + u_3' \cos t = 0$$
$$u_1' e^t - u_2' \cos t - u_3' \sin t = g(t)$$

Solving these equations using either determinants or by elimination, we obtain $u_1' = (1/2)e^{-t}g(t)$,

$u_2' = (1/2)(\sin t - \cos t)g(t)$, $u_3' = -(1/2)(\sin t + \cos t)g(t)$. Integrating these and substituting into Y yields

$$Y(t) = \frac{1}{2}\{e^t \int_{t_0}^t e^{-s}g(s)\,ds + \cos t \int_{t_0}^t (\sin s - \cos s)g(s)\,ds$$
$$-\sin t \int_{t_0}^t (\sin s + \cos s)g(s)\,ds\}.$$

This can be written in the form

$$Y(t) = (1/2)\int_{t_0}^t (e^{t-s} + \cos t \sin s - \cos t \cos s$$
$$-\sin t \sin s - \sin t \cos s)g(s)\,ds.$$

If we use the trigonometric identities $\sin(A-B) = \sin A \cos B - \cos A \sin B$ and $\cos(A-B) = \cos A \cos B + \sin A \sin B$, we obtain the desired result. Note: Eqs.(11) and (12) of this section give the same result, but it is not recommended to memorize these equations.

16. The particular solution has the form $Y = e^t u_1(t) + te^t u_2(t) + t^2 e^t u_3(t)$. Differentiating, making the same assumptions as in the earlier problems, and solving the three linear equations for u_1', u_2', and u_3' yields

$u_1' = (1/2)t^2 e^{-t}g(t)$, $u_2' = -te^{-t}g(t)$ and $u_3' = (1/2)e^{-t}g(t)$. Integrating and substituting into Y yields the desired solution. For instance

$$te^t u_2 = -te^t \int_{t_0}^t s e^{-s}g(s)\,ds = -\frac{1}{2}\int_{t_0}^t 2ts e^{(t-s)}g(s)\,ds, \text{ and}$$

likewise for u_1 and u_3. If $g(t) = t^{-2}e^t$ then $g(s) = e^s/s^2$ and the integration is accomplished using the power rule. Note that terms involving t_0 become part of the complimentary solution.

CHAPTER 5

Section 5.1, Page 231

2. Use the ratio test:
$$\lim_{n \to \infty} \frac{\left| (n+1) x^{n+1}/2^{n+1} \right|}{\left| nx^n/2^n \right|} = \lim_{n \to \infty} \frac{n+1}{n} \frac{1}{2} |x| = \frac{|x|}{2}.$$
Therefore the series converges absolutely for $|x| < 2$. For $x = 2$ and $x = -2$ the nth term does not approach zero as $n \to \infty$ so the series diverge. Hence the radius of convergence is $\rho = 2$.

5. Use the ratio test:
$$\lim_{n \to \infty} \frac{\left| (2x+1)^{n+1}/ (n+1)^2 \right|}{\left| (2x+1)^n/n^2 \right|} = \lim_{n \to \infty} \frac{n^2}{(n+1)^2} |2x+1| = |2x+1|.$$
Therefore the series converges absolutely for $|x+1/2| < 1/2$. At $x = 0$ and $x = -1$ the series also converge absolutely. However, for $|x+1/2| > 1/2$ the series diverges by the ratio test. The radius of convergence is $\rho = 1/2$.

9. For this problem $f(x) = \sin x$. Hence $f'(x) = \cos x$, $f''(x) = -\sin x$, $f'''(x) = -\cos x, \ldots$. Then $f(0) = 0$, $f'(0) = 1$, $f''(0) = 0$, $f'''(0) = -1, \ldots$. The even terms in the series will vanish and the odd terms will alternate in sign. We obtain $\sin x = \sum_{n=0}^{\infty} (-1)^n x^{2n+1}/(2n+1)!$. From the ratio test it follows that $\rho = \infty$.

12. For this problem $f(x) = x^2$. Hence $f'(x) = 2x$, $f''(x) = 2$, and $f^{(n)}(x) = 0$ for $n > 2$. Then $f(-1) = 1$, $f'(-1) = -2$, $f''(-1) = 2$ and $x^2 = 1 - 2(x+1) + 2(x+1)^2/2! = 1 - 2(x+1) + (x+1)^2$. Since the series terminates after a finite number of terms, it converges for all x. Thus $\rho = \infty$.

13. For this problem $f(x) = \ln x$. Hence $f'(x) = 1/x$, $f''(x) = -1/x^2$, $f'''(x) = 1\cdot2/x^3, \ldots$, and $f^{(n)}(x) = (-1)^{n+1}(n-1)!/x^n$. Then $f(1) = 0$, $f'(1) = 1$, $f''(1) = -1$, $f'''(1) = 1\cdot2, \ldots$, $f^{(n)}(1) = (-1)^{n+1}(n-1)!$ The Taylor series is $\ln x = (x-1) - (x-1)^2/2 + (x-1)^3/3 - \ldots = \sum_{n=1}^{\infty} (-1)^{n+1}(x-1)^n/n$. It follows from the ratio test that

the series converges absolutely for $|x-1| < 1$. However, the series diverges at $x = 0$ so $\rho = 1$.

19. Set $m = n-1$ on the right hand side of the equation. Then $n = m+1$ and when $n = 1$, $m = 0$. Thus the right hand side becomes $\displaystyle\sum_{m=0}^{\infty} a_m(x-1)^{m+1}$, which is the same as the left hand side when m is replaced by n.

23. Multiplying each term of the first series by x yields

$$x\sum_{n=1}^{\infty} na_n x^{n-1} = \sum_{n=1}^{\infty} na_n x^n = \sum_{n=0}^{\infty} na_n x^n,\text{ where the last}$$

equality can be verified by writing out the first few terms. Changing the index from k to n ($n=k$) in the second series then yields

$$\sum_{n=0}^{\infty} na_n x^n + \sum_{n=0}^{\infty} a_n x^n = \sum_{n=0}^{\infty} (n+1) a_n x^n.$$

25. $\displaystyle\sum_{m=2}^{\infty} m(m-1) a_m x^{m-2} + x\sum_{k=1}^{\infty} ka_k x^{k-1} =$

$$\sum_{n=0}^{\infty} (n+2)(n+1) a_{n+2} x^n + \sum_{k=1}^{\infty} ka_k x^k =$$

$\displaystyle\sum_{n=0}^{\infty} [(n+2)(n+1) a_{n+2} + na_n]x^n$. In the first case we have

let $n = m - 2$ in the first summation and multiplied each term of the second summation by x. In the second case we have let $n = k$ and noted that for $n = 0$, $na_n = 0$.

28. If we shift the index of summation in the first sum by letting $m = n-1$, we have

$$\sum_{n=1}^{\infty} n\, a_n x^{n-1} = \sum_{m=0}^{\infty} (m+1)\, a_{m+1} x^m.\text{ Substituting this into the}$$

given equation and letting $m = n$ again, we obtain:

$$\sum_{n=0}^{\infty} (n+1) a_{n+1}\, x^n + 2\sum_{n=0}^{\infty} a_n x^n = 0,\text{ or}$$

$$\sum_{n=0}^{\infty} [(n+1)a_{n+1} + 2a_n]x^n = 0.$$

Hence $a_{n+1} = -2a_n/(n+1)$ for $n = 0,1,2,3,\ldots$. Thus

$a_1 = -2a_0$, $a_2 = -2a_1/2 = 2^2a_0/2$, $a_3 = -2a_2/3 = -2^3a_0/2\cdot3 = -2^3a_0/3!\ldots$ and $a_n = (-1)^n2^na_0/n!$. Notice that for $n = 0$ this formula reduces to a_0 so we can write

$$\sum_{n=0}^{\infty} a_nx^n = \sum_{n=0}^{\infty} (-1)^n2^n \, a_0x^n/n! = a_0\sum_{n=0}^{\infty} (-2x)^n/n! = a_0e^{-2x}.$$

Section 5.2, Page 241

2. $y = \displaystyle\sum_{n=0}^{\infty} a_nx^n$; $y' = \displaystyle\sum_{n=1}^{\infty} na_nx^{n-1}$ and since we must multiply

y' by x in the D.E. we do not shift the index; and

$y'' = \displaystyle\sum_{n=2}^{\infty} n(n-1)a_nx^{n-2} = \displaystyle\sum_{n=0}^{\infty} (n+2)(n+1)a_{n+2}x^n$. Substituting

in the D.E., we obtain

$$\sum_{n=0}^{\infty} (n+2)(n+1)a_{n+2}x^n - \sum_{n=1}^{\infty} na_nx^n - \sum_{n=0}^{\infty} a_nx^n = 0.$$ In order to

have the starting point the same in all three summations, we let $n = 0$ in the first and third terms to obtain the following

$$(2\cdot1 \, a_2 - a_0)x^0 + \sum_{n=1}^{\infty} [(n+2)(n+1)a_{n+2} - (n+1)a_n]x^n = 0.$$

Thus $a_{n+2} = a_n/(n+2)$ for $n = 1,2,3,\ldots$. Note that the recurrence relation is also correct for $n = 0$. We show how to calculate the odd a's:

$a_3 = a_1/3$, $a_5 = a_3/5 = a_1/5\cdot3$, $a_7 = a_5/7 = a_1/7\cdot5\cdot3$, $\ldots$.

Now notice that $a_3 = 2a_1/(2\cdot3) = 2a_1/3!$, that

$a_5 = 2\cdot4a_1/(2\cdot3\cdot4\cdot5) = 2^2 2a_1/5!$, and that

$a_7 = 2\cdot4\cdot6a_1/(2\cdot3\cdot4\cdot5\cdot6\cdot7) = 2^3 3! \, a_1/7!$. Likewise

$a_9 = a_7/9 = 2^3 3! \, a_1/(7!)9 = 2^3 3! \, 8a_1/9! = 2^4 4! \, a_1/9!$.

Continuing we have $a_{2m+1} = 2^m m! \, a_1/(2m+1)!$. In the same way we find that the even a's are given by $a_{2m} = a_0/2^m \, m!$.

Thus $\quad y = a_0 \sum_{m=0}^{\infty} \dfrac{x^{2m}}{2^m m!} + a_1 \sum_{m=0}^{\infty} \dfrac{2^m m! \; x^{2m+1}}{(2m+1)!}$.

3. $\quad y = \sum_{n=0}^{\infty} a_n (x-1)^n; \; y' = \sum_{n=1}^{\infty} n a_n (x-1)^{n-1} = \sum_{n=0}^{\infty} (n+1) a_{n+1} (x-1)^n$,

and

$y'' = \sum_{n=2}^{\infty} n(n-1) a_n (x-1)^{n-2} = \sum_{n=0}^{\infty} (n+2)(n+1) a_{n+2} (x-1)^n$.

Substituting in the D.E. and setting $x = 1 + (x-1)$ we obtain

$$\sum_{n=0}^{\infty} (n+2)(n+1) a_{n+2} (x-1)^n - \sum_{n=0}^{\infty} (n+1) a_{n+1} (x-1)^n - \sum_{n=1}^{\infty} n a_n (x-1)^n$$

$$- \sum_{n=0}^{\infty} a_n (x-1)^n = 0.$$

Letting $n = 0$ in the first, second, and the fourth sums, we obtain

$$(2 \cdot 1 \cdot a_2 - 1 \cdot a_1 - a_0)(x-1)^0 + \sum_{n=1}^{\infty} [(n+2)(n+1) a_{n+2}$$

$$- (n+1) a_{n+1} - (n+1) a_n](x-1)^n = 0.$$

Thus $(n+2) a_{n+2} - a_{n+1} - a_n = 0$ for $n = 0, 1, 2, \ldots$. This recurrance relation can be used to solve for a_2 in terms of a_0 and a_1, then for a_3 in terms of a_0 and a_1, etc. In many cases it is easier to first take $a_0 = 0$ and generate one solution and then take $a_1 = 0$ and generate the second linearly independent solution. Thus, choosing $a_0 = 0$ we find that $a_2 = a_1/2$, $a_3 = (a_2+a_1)/3 = a_1/2$, $a_4 = (a_3+a_2)/4 = a_1/4$, $a_5 = (a_4+a_3)/5 = 3a_1/20, \ldots$. This yields the solution $y_2(x) = a_1[(x-1) + (x-1)^2/2 + (x-1)^3/2 + (x-1)^4/4 + 3(x-1)^5/20 + \ldots]$. The second independent solution may be obtained by choosing $a_1 = 0$. Then $a_2 = a_0/2$, $a_3 = (a_2+a_1)/3 = a_0/6$, $a_4 = (a_3+a_2)/4 = a_0/6$, $a_5 = (a_4+a_3)/5 = a_0/15, \ldots$. This yields the solution $y_1(x) = a_0[1 + (x-1)^2/2 + (x-1)^3/6 + (x-1)^4/6 + (x-1)^5/15 + \ldots]$.

5. $y = \sum_{n=0}^{\infty} a_n x^n$; $y' = \sum_{n=1}^{\infty} n a_n x^{n-1}$; and $y'' = \sum_{n=2}^{\infty} n(n-1) a_n x^{n-2}$.

Substituting in the D.E. and shifting the index in both summations for y'' gives

$$\sum_{n=0}^{\infty} (n+2)(n+1) a_{n+2} x^n - \sum_{n=1}^{\infty} (n+1) n \ a_{n+1} x^n + \sum_{n=0}^{\infty} a_n x^n =$$

$$(2 \cdot 1 \cdot a_2 + a_0) x^0 + \sum_{n=1}^{\infty} [(n+2)(n+1) a_{n+2} - (n+1) n a_{n+1} + a_n] x^n = 0.$$

Thus $a_2 = -a_0/2$ and $a_{n+2} = n a_{n+1}/(n+2) - a_n/(n+2)(n+1)$, $n = 1, 2, \ldots$. Choosing $a_0 = 0$ yields $a_2 = 0$, $a_3 = -a_1/6$, $a_4 = 2a_3/4 = -a_1/12, \ldots$ which gives one solution as $y_2(x) = a_1(x - x^3/6 - x^4/12 + \ldots)$. A second linearly independent solution is obtained by choosing $a_1 = 0$. Then $a_2 = -a_0/2$, $a_3 = a_2/3 = -a_0/6$, $a_4 = 2a_3/4 - a_2/12 = -a_0/24, \ldots$ which gives $y_1(x) = a_0(1 - x^2/2 - x^3/6 - x^4/24 + \ldots)$.

14. You will need to rewrite $x+1$ as $3 + (x-2)$ in order to multiply $x+1$ times y' as a power series about $x_0 = 2$.

16a. From Problem 6 we have

$$y(x) = c_1(1 - x^2 + \frac{1}{6} x^4 + \ldots) + c_2(x - \frac{1}{4} x^3 + \frac{7}{160} x^5 + \ldots).$$

Now $y(0) = c_1 = -1$ and $y'(0) = c_2 = 3$ and thus

$$y(x) = -1 + x^2 - \frac{1}{6} x^4 + 3x - \frac{3}{4} x^3 = -1 + 3x + x^2 - \frac{3}{4} x^3 - \frac{1}{6} x^4 + \ldots .$$

16c. By plotting $f = -1 + 3x + x^2 - 3x^3/4$ and $g = f - x^4/6$ between -1 and 1 it appears that f is a reasonable approximation for $|x| < 0.7$.

19. The D.E. transforms into $u''(t) + t^2 u'(t) + (t^2 + 2t) u(t) = 0$.

Assuming that $u(t) = \sum_{n=0}^{\infty} a_n t^n$, we have $u'(t) = \sum_{n=1}^{\infty} n a_n t^{n-1}$ and

$u''(t) = \sum_{n=2}^{\infty} n(n-1) a_n t^{n-2}$. Substituting in the D.E. and shifting indices yields

$$\sum_{n=0}^{\infty} (n+2)(n+1) a_{n+2}t^n + \sum_{n=2}^{\infty} (n-1) a_{n-1}t^n + \sum_{n=2}^{\infty} a_{n-2}t^n$$

$$+ \sum_{n=1}^{\infty} 2a_{n-1}t^n = 0,$$

$$2\cdot1\cdot a_2 t^0 + (3\cdot2\cdot a_3 + 2\cdot a_0)t^1 + \sum_{n=2}^{\infty} [(n+2)(n+1)a_{n+2}$$

$$+ (n+1)a_{n-1} + a_{n-2}]t^n = 0.$$

It follows that $a_2 = 0$, $a_3 = -a_0/3$ and
$a_{n+2} = -a_{n-1}/(n+2) - a_{n-2}/[(n+2)(n+1)]$, $n = 2,3,4...$. We
obtain one solution by choosing $a_1 = 0$. Then $a_4 = -a_0/12$,
$a_5 = -a_2/5 - a_1/20 = 0$, $a_6 = -a_3/6 - a_2/30 = a_0/18,...$. Thus
one solution is $u_1(t) = a_0(1 - t^3/3 - t^4/12 + t^6/18 + ...)$ so
$y_1(x) = u_1(x-1) = a_0[1 - (x-1)^3/3 - (x-1)^4/12 + (x-1)^6/18 + ...]$.
We obtain a second solution by choosing $a_0 = 0$. Then
$a_4 = -a_1/4$, $a_5 = -a_2/5 - a_1/20 = -a_1/20$,
$a_6 = -a_3/6 - a_2/30 = 0$, $a_7 = -a_4/7 - a_3/42 = a_1/28,...$.
Thus a second linearly independent solution is
$u_2(t) = a_1[t - t^4/4 - t^5/20 + t^7/28 + ...]$ or
$y_2(x) = u_2(x-1)$

$$= a_1[(x-1) - (x-1)^4/4 - (x-1)^5/20 + (x-1)^7/28 + ...].$$

The Taylor series for $x^2 - 1$ about $x = 1$ may be obtained by
writing $x = (x-1) + 1$ so $x^2 = (x-1)^2 + 2(x-1) + 1$ and
$x^2 - 1 = (x-1)^2 + 2(x-1)$. The D.E. now appears as
$y'' + (x-1)^2 y' + [(x-1)^2 + 2(x-1)]y = 0$ which is identical to
the transformed equation with $t = x - 1$.

22b. $y = a_0 + a_1 x + a_2 x^2 + ...$, $y^2 = a_0^2 + 2a_0 a_1 x + (2a_0 a_2 + a_1^2)x^2$
$+ ...$, $y' = a_1 + 2a_2 x + 3a_3 x^2 + ...$, and
$(y')^2 = a_1^2 + 4a_1 a_2 x + (6a_1 a_3 + 4a_2^2)x^2 + ...$. Substituting
these into $(y')^2 = 1 - y^2$ and collecting coefficients of
like powers of x yields $(a_1^2 + a_0^2 - 1) + (4a_1 a_2 + 2a_0 a_1)x +$
$(6a_1 a_3 + 4a_2^2 + 2a_0 a_2 + a_1^2)x^2 + ... = 0$. As in the earlier
problems, each coefficient must be zero. The I.C. $y(0) = 0$
requires that $a_0 = 0$, and thus $a_1^2 + a_0^2 - 1 = 0$ gives $a_1^2 = 1$.
However, the D.E.indicates that y' is always positive, so

$y'(0) = a_1 > 0$ implies $a_1 = 1$. Then $4a_1a_2 + 2a_0a_1 = 0$ implies that $a_2 = 0$; and $6a_1a_3 + 4a_2^2 + 2a_0a_2 + a_1^2 = 6a_1a_3 + a_1^2 = 0$ implies that $a_3 = -1/6$. Thus $y = x - x^3/3! + \ldots$.

23.   26.

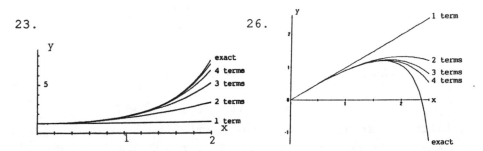

26. We have $y(x) = a_0y_1 + a_1y_2$, where y_1 and y_2 are found in Problem 10. Now $y(0) = a_0 = 0$ and $y'(0) = a_1 = 1$. Thus

$$y(x) = x - \frac{x^3}{12} - \frac{x^5}{240} - \frac{x^7}{2240}.$$

Section 5.3, Page 247

1. The D.E. can be solved for y'' to yield $y'' = -xy' - y$. If $y = \phi(x)$ is a solution, then $\phi''(x) = -x\phi'(x) - \phi(x)$ and thus setting $x = 0$ we obtain $\phi''(0) = -0 - 1 = -1$. Differentiating the equation for y'' yields $y''' = -xy'' - 2y'$ and hence setting $y = \phi(x)$ again yields $\phi'''(0) = -0 - 0 = 0$. In a similar fashion $y^{iv} = -xy''' - 3y''$ and thus $\phi^{iv}(0) = -0 - 3(-1) = 3$. The process can be continued to calculate higher derivatives of $\phi(x)$.

6. The zeros of $P(x) = x^2 - 2x - 3$ are $x = -1$ and $x = 3$. For $x_0 = 4$, $x_0 = -4$, and $x_0 = 0$ the distance to the nearest zero of $P(x)$ is 1,3, and 1, respectively. Thus a lower bound for the radius of convergence for series solutions in powers of $(x-4)$, $(x+4)$, and x is $\rho = 1$, $\rho = 3$, and $\rho = 1$, respectively.

9a. Since $P(x) = 1$ has no zeros, the radius of convergence is $\rho = \infty$.

9f. Since $P(x) = x^2 + 2$ has zeros at $x = \pm\sqrt{2}\ i$, the lower bound for the radius of convergence of the series solution about $x_0 = 0$ is $\rho = \sqrt{2}$.

10a. If we assume that $y = \sum_{n=0}^{\infty} a_n x^n$, then $y' = \sum_{n=1}^{\infty} n a_n x^{n-1}$ and

$y'' = \sum_{n=2}^{\infty} n(n-1) a_n x^{n-2}$. Substituting in the D.E., shifting

indices of summation, and collecting coefficients of like powers of x yields the equation

$(2 \cdot 1 \cdot a_2 + \alpha^2 a_0) x^0 + [3 \cdot 2 \cdot a_3 + (\alpha^2 - 1) a_1] x^1$

$+ \sum_{n=2}^{\infty} [(n+2)(n+1) a_{n+2} + (\alpha^2 - n^2) a_n] x^n = 0$.

Hence the recurrence relation is
$a_{n+2} = (n^2 - \alpha^2) a_n / (n+2)(n+1)$, $n = 0, 1, 2, \ldots$. For the first solution we choose $a_1 = 0$. We find that
$a_2 = -\alpha^2 a_0 / 2 \cdot 1$, $a_3 = 0$, $a_4 = (2^2 - \alpha^2) a_2 / 4 \cdot 3 = -(2^2 - \alpha^2) \alpha^2 a_0 / 4!$
$\ldots$, $a_{2m} = -[(2m-2)^2 - \alpha^2] \ldots (2^2 - \alpha^2) \alpha^2 a_0 / (2m)!$,

and $a_{2m+1} = 0$, so $y_1(x) = 1 - \dfrac{\alpha^2}{2!} x^2 - \dfrac{(2^2 - \alpha^2) \alpha^2}{4!} x^4 - \ldots$

$- \dfrac{[(2m-2)^2 - \alpha^2] \ldots (2^2 - \alpha^2) \alpha^2}{(2m)!} x^{2m} - \ldots$

where we have set $a_0 = 1$. For the second solution we take $a_0 = 0$ and $a_1 = 1$ in the recurrence relation to obtain the desired solution.

10b. If α is an even integer $2k$ then $(2m-2)^2 - \alpha^2 = (2m-2)^2 - 4k^2 = 0$. Thus when $m = k+1$ all terms in the series for $y_1(x)$ are zero after the x^{2k} term. A similar argument shows that if $\alpha = 2k+1$ then all terms in $y_2(x)$ are zero after the x^{2k+1}.

11. The Taylor series about $x = 0$ for $\sin x$ is
$\sin x = x - x^3/3! + x^5/5! - \ldots$. Assuming that

$y = \sum_{n=0}^{\infty} a_n x^n$ we find $y'' + (\sin x) y = 2a_2 + 6a_3 x + 12a_4 x^2$

$+ 20a_5 x^3 + 30a_6 x^4 + 42a_7 x^5 + \ldots$

$+ (x - x^3/3! + x^5/5! - \ldots)(a_0 + a_1 x + a_2 x^2 + a_3 x^3 + a_4 x^4 + \ldots)$

$= 2a_2 + (6a_3 + a_0) x + (12a_4 + a_1) x^2 + (20a_5 + a_2 - a_0/6) x^3 +$

$(30a_6 + a_3 - a_1/6) x^4 + (42a_7 + a_4 + a_0/120) x^5 + \ldots = 0$. Hence

$a_2 = 0$, $a_3 = -a_0/6$, $a_4 = -a_1/12$, $a_5 = a_0/120$,
$a_6 = (a_1+a_0)/180$, $a_7 = a_0/7! + a_1/504$, ... We set
$a_0 = 1$ and $a_1 = 0$ and obtain
$y_1(x) = (1 - x^3/6 + x^5/120 + x^6/180 + ...)$. Next we set
$a_0 = 0$ and $a_1 = 1$ and obtain
$y_2(x) = (x - x^4/12 + x^6/180 + x^7/504 + ...)$. Since
$p(x) = 1$ and $q(x) = \sin x$ both have $\rho = \infty$, the solution
in this case converges for all x, that is, $\rho = \infty$

20. Substituting $y = \sum_{n=0}^{\infty} a_n x^n$ into the D.E. we obtain

$\sum_{n=1}^{\infty} n a_n x^{n-1} - \sum_{n=0}^{\infty} a_n x^n = x^2$. Shifting indices in the

summation yields $\sum_{n=0}^{\infty} [(n+1)a_{n+1} - a_n] x^n = x^2$. Equating

coefficients of both sides then gives: $a_1 - a_0 = 0$,
$2a_2 - a_1 = 0$, $3a_3 - a_2 = 1$ and $(n+1)a_{n+1} = a_n$ for
$n = 3,4,...$. Thus $a_1 = a_0$, $a_2 = a_1/2 = a_0/2$, $a_3 = 1/3 + a_2/3 = 1/3 + a_0/2\cdot3$, $a_4 = a_3/4 = 1/3\cdot4 + a_0/2\cdot3\cdot4$, ...,
$a_n = a_{n-1}/n = 2/n! + a_0/n!$ and hence

$$y(x) = a_0(1 + x + \frac{x^2}{2!} + ... + \frac{x^n}{n!} ...) + 2(\frac{x^3}{3!} + \frac{x^4}{4!} + ... + \frac{x^n}{n!} + ...).$$

Using the power series for e^x, the first and second sums
can be rewritten as $a_0 e^x + 2(e^x - 1 - x - x^2/2)$.

22. Substituting $y = \sum_{n=0}^{\infty} a_n x^n$ into the Legendre equation,

shifting indices, and collecting coefficients of like
powers of x yields
$[2\cdot1\cdot a_2 + \alpha(\alpha+1)a_0]x^0 + \{3\cdot2\cdot a_3 - [2\cdot1 - \alpha(\alpha+1)]a_1\}x^1 +$

$\sum_{n=2}^{\infty} \{(n+2)(n+1)a_{n+2} - [n(n+1) - \alpha(\alpha+1)]a_n\}x^n = 0$. Thus

$a_2 = -\alpha(\alpha+1)a_0/2!$, $a_3 = [2\cdot1 - \alpha(\alpha+1)]a_1/3! = -(\alpha-1)(\alpha+2)a_1/3!$ and the recurrence relation is
$(n+2)(n+1)a_{n+2} = -[\alpha(\alpha+1) - n(n+1)]a_n = -(\alpha-n)(\alpha+n+1)a_n$,
$n = 2,3,...$. Setting $a_1 = 0$, $a_0 = 1$ yields a solution

with $a_3 = a_5 = a_7 = \ldots = 0$ and

$a_4 = \alpha(\alpha-2)(\alpha+1)(\alpha+3)/4!,\ldots, \; a_{2m} = (-1)^m\alpha(\alpha-2)(\alpha-4)\ldots$
$(\alpha-2m+2)(\alpha+1)(\alpha+3)\ldots(\alpha+2m-1)/(2m)!,\ldots$. The second
linearly independent solution is obtained by setting
$a_0 = 0$ and $a_1 = 1$. The coefficients are $a_2 = a_4 = a_6 =$
$\ldots = 0$ and $a_3 = -(\alpha-1)(\alpha+2)/3!$, $a_5 = -(\alpha-3)(\alpha+4)a_3/5\cdot4 =$
$(\alpha-1)(\alpha-3)(\alpha+2)(\alpha+4)/5!,\ldots$.

26. Using the chain rule we have:
$$\frac{dF(\phi)}{d\phi} = \frac{dF[\phi(x)]}{dx}\frac{dx}{d\phi} = -f'(x)\sin\phi(x) = -f'(x)\sqrt{1-x^2},$$
$$\frac{d^2F(\phi)}{d\phi^2} = \frac{d}{dx}[-f'(x)\sqrt{1-x^2}]\frac{dx}{d\phi} = (1-x^2)f''(x) - xf'(x),$$
which when substituted into the D.E. yields the desired
result.

28. Carrying out the steps indicated yields the two
equations:
$$P_m[(1-x^2)P_n']' = -n(n+1)P_nP_m$$
$$P_n[(1-x^2)P_m']' = -m(m+1)P_nP_m.$$
As long as $n \neq m$ the second equation can be subtracted
from the first and the result integrated from -1 to 1 to
obtain
$$\int_{-1}^{1}\{P_m[(1-x^2)P_n']'-P_n[(1-x^2)P_m']'\}dx = [m(m+1)-n(n+1)]\int_{-1}^{1}P_nP_mdx$$
The left side may be integrated by parts to yield
$$[P_m(1-x^2)P_n' - P_n(1-x^2)P_m']_{-1}^{1} + \int_{-1}^{1}[P_m'(1-x^2)P_n' - P_n'(1-x^2)P_m']dx,$$
which is zero. Thus $\int_{-1}^{1}P_n(x)P_m(x)dx = 0$ for $n \neq m$.

Section 5.4, Page 253

1. Since the coefficients of y, y' and y'' have no common
factors and since $P(x)$ vanishes only at $x = 0$ we conclude
that $x = 0$ is a singular point. Writing the D.E. in the
form $y'' + p(x)y' + q(x)y = 0$, we obtain $p(x) = (1-x)/x$
and $q(x) = 1$. Thus for the singular point we have
$\lim_{x\to0} x\,p(x) = \lim_{x\to0} 1-x = 1$, $\lim_{x\to0} x^2q(x) = 0$ and thus $x = 0$
is a regular singular point.

5. Writing the D.E. in the form $y'' + p(x)y' + q(x)y = 0$, we
find $p(x) = x/(1-x)(1+x)^2$ and $q(x) = 1/(1-x^2)(1+x)$.

Therefore $x = \pm 1$ are singular points. Since $\lim\limits_{x \to 1} (x-1)p(x)$ and $\lim\limits_{x \to 1} (x-1)^2 q(x)$ both exist, we conclude $x = 1$ is a regular singular point. Finally, since $\lim\limits_{x \to -1} (x+1)p(x)$ does not exist, we find that $x = -1$ is an irregular singular point.

12. Writing the D.E. in the form $y + p(x)y' + q(x)y = 0$, we see that $p(x) = e^x/x$ and $q(x) = (3\cos x)/x$. Thus $x = 0$ is a singular point. Since $xp(x) = e^x$ is analytic at $x = 0$ and $x^2 q(x) = 3x\cos x$ is analytic at $x = 0$ the point $x = 0$ is a regular singular point.

17. Writing the D.E. in the form $y'' + p(x)y' + q(x)y = 0$, we see that $p(x) = \dfrac{x}{\sin x}$ and $q(x) = \dfrac{4}{\sin x}$. Since $\lim\limits_{x \to 0} q(x)$ does not exist, the point $x_0 = 0$ is a singular point and since neither $\lim\limits_{x \to \pm n\pi} p(x)$ nor $\lim\limits_{x \to \pm n\pi} q(x)$ exist either the points $x_0 = \pm n\pi$ are also singular points. To determine whether the singular points are regular or irregular we must use Eq.(8) and the result #7 of multiplication and division of power series from Section 5.1. For $x_0 = 0$, we have

$$xp(x) = \frac{x^2}{\sin x} = \frac{x^2}{x - \dfrac{x^3}{6} + \ldots} = x[1 + \frac{x^2}{6} + \ldots]$$

$$= x + \frac{x^3}{6} + \ldots,$$

which converges about $x_0 = 0$ and thus $xp(x)$ is analytic at $x_0 = 0$. $x^2 q(x)$, by similar steps, is also analytic at $x_0 = 0$ and thus $x_0 = 0$ is a regular singular point. For $x_0 = n\pi$, we have

$$(x-n\pi)p(x) = \frac{(x-n\pi)x}{\sin x} = \frac{(x-n\pi)[(x-n\pi) + n\pi]}{\pm(x-n\pi) + \dfrac{-(x-n\pi)^3}{6} \pm \ldots}$$

$$= [(x-n\pi) + n\pi][\pm 1 \pm \frac{(x-n\pi)^2}{6} \pm \ldots], \text{ which}$$

converges about $x_0 = n\pi$ and thus $(x-n\pi)p(x)$ is analytic at $x = n\pi$. Similarly $(x+n\pi)p(x)$ and $(x\pm n\pi)^2 q(x)$ are analytic and thus $x_0 = \pm n\pi$ are regular singular points.

19. Substituting $y = \sum_{n=0}^{\infty} a_n x^n$ into the D.E. yields

$2\sum_{n=2}^{\infty} n(n-1) a_n x^{n-1} + 3\sum_{n=1}^{\infty} n a_n x^{n-1} + \sum_{n=0}^{\infty} a_n x^{n+1} = 0$.   The last sum

becomes $\sum_{n=2}^{\infty} a_{n-2} x^{n-1}$ by replacing n+1 by n-1, the first term

of the middle sum is $3a_1$, and thus we have

$3a_1 + \sum_{n=2}^{\infty} \{[2n(n-1)+3n]a_n + a_{n-2}\} x^{n-1} = 0$.   Hence $a_1 = 0$ and

$a_n = \dfrac{-a_{n-2}}{n(2n+1)}$, which is the desired recurrance relation.

Thus all even coefficients are found in terms of a_0 and
all odd coefficients are zero, thereby yielding only one
solution of the desired form.

21. If $\xi = 1/x$ then

$\dfrac{dy}{dx} = \dfrac{dy}{d\xi} \dfrac{d\xi}{dx} = -\dfrac{1}{x^2} \dfrac{dy}{d\xi} = -\xi^2 \dfrac{dy}{d\xi}$,

$\dfrac{d^2y}{dx^2} = \dfrac{d}{d\xi}(-\xi^2 \dfrac{dy}{d\xi}) \dfrac{d\xi}{dx} = (-2\xi \dfrac{dy}{d\xi} - \xi^2 \dfrac{d^2y}{d\xi^2}) (-\dfrac{1}{x^2})$

$= \xi^4 \dfrac{d^2y}{d\xi^2} + 2\xi^3 \dfrac{dy}{d\xi}$.

Substituting in the D.E. we have

$P(1/\xi) [\xi^4 \dfrac{d^2y}{d\xi^2} + 2\xi^3 \dfrac{dy}{d\xi}] + Q(1/\xi) [-\xi^2 \dfrac{dy}{d\xi}] + R(1/\xi)y = 0$,

$\xi^4 P(1/\xi) \dfrac{d^2y}{d\xi^2} + [2\xi^3 P(1/\xi) - \xi^2 Q(1/\xi)] \dfrac{dy}{d\xi} + R(1/\xi)y = 0$.

The result then follows from the theory of singular
points at $\xi = 0$.

23. Since $P(x) = x^2$, $Q(x) = x$ and $R(x) = -4$ we have
$f(\xi) = [2P(1/\xi)/\xi - Q(1/\xi)/\xi^2]/P(1/\xi) = 2/\xi - 1/\xi = 1/\xi$
and $g(\xi) = R(1/\xi)/\xi^4 P(1/\xi) = -4/\xi^2$.   Thus the point at
infinity is a singular point.   Since both $\xi f(\xi)$ and
$\xi^2 g(\xi)$ are analytic at $\xi = 0$, the point at infinity is a
regular singular point.

25. Since $P(x) = x^2$, $Q(x) = x$, and $R(x) = x^2 - v^2$,
     $f(\xi) = [2P(1/\xi)/\xi - Q(1/\xi)/\xi^2]/P(1/\xi) = 2/\xi - 1/\xi = 1/\xi$
     and $g(\xi) = R(1/\xi)/\xi^4 P(1/\xi) = (1/\xi^2 - v^2)/\xi^2 = 1/\xi^4 - v^2/\xi^2$.
     Thus the point at infinity is a singular point. Although
     $\xi f(\xi) = 1$ is analytic at $\xi = 0$, $\xi^2 g(\xi) = 1/\xi^2 - v^2$ is not,
     so the point at infinity is an irregular singular point.

Section 5.5, Page 260

2.   Assume $y = (x+1)^r$ for $x + 1 > 0$. Substitution of y into
     the D.E. yields $[r(r-1) + 3r + 3/4](x+1)^r = 0$. Thus
     $r^2 + 2r + 3/4 = 0$, which yields $r = -3/2, -1/2$. The
     general solution of the D.E. is then
     $y = c_1|x+1|^{-1/2} + c_2|x+1|^{-3/2}$, $x \neq -1$.

4.   If $y = x^r$ then $r(r-1) + 3r + 5 = 0$. So $r^2 + 2r + 5 = 0$
     and $r = (-2 \pm \sqrt{4-20})/2 = -1 \pm 2i$. Thus the general
     solution of the D.E. is
     $y = c_1 x^{-1}\cos(2\ln|x|) + c_2 x^{-1}\sin(2\ln|x|)$, $x \neq 0$.

9.   Again let $y = x^r$ to obtain $r(r-1) - 5r + 9 = 0$, or
     $(r-3)^2 = 0$. Thus the roots are $x = 3,3$ and
     $y = c_1 x^3 + c_2 x^3 \ln|x|$, $x \neq 0$, is the solution of the D.E.

13. If $y = x^r$, then $F(r) = 2r(r-1) + r -3 = 2r^2 - r - 3 =$
     $(2r-3)(r+1) = 0$, so $y = c_1 x^{3/2} + c_2 x^{-1}$ and

     $y' = \frac{3}{2}c_1 x^{1/2} - c_2 x^{-2}$. Setting $x = 1$ in y and y' we obtain

     $c_1 + c_2 = 1$ and $\frac{3}{2}c_1 - c_2 = 4$, which yield $c_1 = 2$ and

     $c_2 = -1$. Hence $y = 2x^{3/2} - x^{-1}$. As $x \to 0^+$ we have
     $y \to -\infty$ due to the second term.

17. Substituting $y = x^r$, we find that $r(r-1) + \alpha r + 5/2 = 0$
     or $r^2 + (\alpha-1)r + 5/2 = 0$. Thus
     $r_1, r_2 = [-(\alpha-1) \pm \sqrt{(\alpha-1)^2-10}]/2$. In order for solutions
     to approach zero as $x \to 0$ it is necessary that the real
     parts of r_1 and r_2 be positive. Suppose that $\alpha > 1$, then
     $\sqrt{(\alpha-1)^2-10}$ is either imaginary or real and less than
     $\alpha - 1$; hence the real parts of r_1 and r_2 will be

negative. Suppose that $\alpha = 1$, then $r_1, r_2 = \pm i\sqrt{10}$ and
the solutions are oscillatory. Suppose that $\alpha < 1$, then
$\sqrt{(\alpha-1)^2-10}$ is either imaginary or real and less than
$|\alpha-1| = 1 - \alpha$; hence the real parts of r_1 and r_2 will be
positive. Thus if $\alpha < 1$ the solutions of the D.E. will
approach zero as $x \to 0$.

21. In all cases the roots of $F(r) = 0$ are given by Eq.(5)
     and the forms of the solution are given in Theorem 5.5.1.

22. Assume that $y = v(x)x^{r_1}$. Then $y' = v(x)r_1x^{r_1-1} + v'(x)x^{r_1}$
     and $y'' = v(x)r_1(r_1-1)x^{r_1-2} + 2v'(x)r_1x^{r_1-1} + v''(x)x^{r_1}$.
     Substituting in the D.E. and collecting terms yields
     $x^{r_1+2}v'' + (\alpha + 2r_1)x^{r_1+1}v' + [r_1(r_1-1) + \alpha r_1 + \beta]x^{r_1}v = 0$.
     Now we make use of the fact that r_1 is a double root of
     $f(r) = r(r-1) + \alpha r + \beta$. This means that $f(r_1) = 0$ and
     $f'(r_1) = 2r_1 - 1 + \alpha = 0$. Hence the D.E. for v reduces
     to $x^{r_1+2}v'' + x^{r_1+1}v'$. Since $x > 0$ we may divide by x^{r_1+1}
     to obtain $xv'' + v' = 0$. Thus $v(x) = \ln x$ and a second
     solution is $y = x^{r_1}\ln x$.

25. The change of variable $x = e^z$ transforms the D.E. into
     $u'' - 4u' + 4u = z$, which has the solution
     $u(z) = c_1e^{2z} + c_2ze^{2z} + (1/4)z + 1/4$. Hence
     $y(x) = c_1x^2 + c_2x^2\ln x + (1/4)\ln x + 1/4$.

31. If $x > 0$, then $|x| = x$ and $|x|^{r_1} = x^{r_1}$ so we can choose
     $c_1 = k_1$. If $x < 0$, then $|x| = -x$ and $|x|^{r_1} = (-x)^{r_1} =$
     $(-1)^{r_1}x^{r_1}$ and we can choose $c_1 = (-1)^{r_1}k_1$, or $k_1 = (-1)^{r_1}c_1$.
     In both bases we have $c_2 = k_2$.

Section 5.6, Page 266

2. If the D.E. is put in the standard form $y'' + p(x)y +$
    $q(x)y = 0$, then $p(x) = x^{-1}$ and $q(x) = 1 - 1/9x^2$. Thus
    $x = 0$ is a singular point. Since $xp(x) \to 1$ and
    $x^2q(x) \to -1/9$ as $x \to 0$ it follows that $x = 0$ is a
    regular singular point. In determining a series solution
    of the D.E. it is more convenient to leave the equation
    in the form given rather than divide by the x^2, the

coefficient of y''. If we substitute $y = \sum_{n=0}^{\infty} a_n x^{n+r}$, we have

$$\sum_{n=0}^{\infty} (n+r)(n+r-1)a_n x^{n+r} + \sum_{n=0}^{\infty} (n+r)a_n x^{n+r} + (x^2 - \frac{1}{9})\sum_{n=0}^{\infty} a_n x^{n+r} = 0.$$

Note that $x^2 \sum_{n=0}^{\infty} a_n x^{n+r} = \sum_{n=0}^{\infty} a_n x^{n+r+2} = \sum_{n=2}^{\infty} a_{n-2} x^{n+r}$. Thus we

have $[r(r-1) + r - \frac{1}{9}]a_0 x^r + [(r+1)r + (r+1) - \frac{1}{9}]a_1 x^{r+1} +$

$$\sum_{n=2}^{\infty} \{[(n+r)(n+r-1) + (n+r) - \frac{1}{9}]a_n + a_{n-2}\} x^{n+r} = 0.$$ The

indicial equation is $r^2 - 1/9 = 0$ with roots $r_1 = 1/3$ and
$r_2 = -1/3$. For either value of r it is necessary to
take $a_1 = 0$ in order that the coefficient of x^{r+1} be zero.
The recurrence relation is $[(n+r)^2 - 1/9]a_n = -a_{n-2}$. For
$r = 1/3$ we have

$$a_n = \frac{-a_{n-2}}{(n + \frac{1}{3})^2 - (\frac{1}{3})^2} = -\frac{a_{n-2}}{(n + \frac{2}{3})n}, \quad n = 2,3,4,\ldots .$$

Since $a_1 = 0$ it follows from the recurrence relation that
$a_3 = a_5 = a_7 = \ldots = 0$. For the even coefficients it is
convenient to let $n = 2m$, $m = 1,2,3,\ldots$. Then
$a_{2m} = -a_{m-2}/2^2 m(m + \frac{1}{3})$. The first few coefficients are
given by

$$a_2 = \frac{(-1)a_0}{2^2(1 + \frac{1}{3})1}, \quad a_4 = \frac{(-1)a_2}{2^2(2 + \frac{1}{3})2} = \frac{a_0}{2^4(1 + \frac{1}{3})(2 + \frac{1}{3})2!}$$

$$a_6 = \frac{(-1)a_4}{2^2(3 + \frac{1}{3})3} = \frac{(-1)a_0}{2^6(1 + \frac{1}{3})(2 + \frac{1}{3})(3 + \frac{1}{3})3!}, \quad \text{and the}$$

coefficient of x^{2m} for $m = 1, 2, \ldots$ is

$$a_{2m} = \frac{(-1)^m a_0}{2^{2m} m!(1 + \frac{1}{3})(2 + \frac{1}{3}) \ldots (m + \frac{1}{3})}. \quad \text{Thus one}$$

solution (on setting $a_0 = 1$) is

$$y_1(x) = x^{1/3}[1 + \sum_{m=1}^{\infty} \frac{(-1)^m}{m! \ (1 + \frac{1}{3})(2 + \frac{1}{3})\ldots(m + \frac{1}{3})} \ (\frac{x}{2})^{2m}].$$

Since $r_2 = -1/3 \neq r_1$ and $r_1 - r_2 = 2/3$ is not an integer, we can calculate a second series solution corresponding to $r = -1/3$. The recurrence relation is $n(n-2/3)a_n = -a_{n-2}$, which yields the desired solution following the steps just outlined. Note that $a_1 = 0$, as in the first solution, and thus all the odd coefficients are zero.

4. Putting the D.E. in standard form $y''+p(x)y'+q(x)y = 0$, we see that $p(x) = 1/x$ and $q(x) = -1/x$. Thus $x = 0$ is a singular point, and since $xp(x) \to 1$ and $x^2q(x) \to 0$, as $x \to 0$, $x = 0$ is a regular singular point. Substituting $y = \sum_{n=0}^{\infty} a_n x^{n+r}$ in $xy'' + y' - y = 0$ and shifting indices we obtain

$$\sum_{n=-1}^{\infty} a_{n+1}(r+n+1)(r+n)x^{n+r} + \sum_{n=-1}^{\infty} a_{n+1}(r+n+1)x^{n+r} - \sum_{n=-}^{\infty} a_n x^{n+r} = 0,$$

$$[r(r-1) + r]a_0 x^{-1+r} + \sum_{n=0}^{\infty} [(r+n+1)^2 a_{n+1} - a_n]x^{n+r} = 0. \quad \text{The}$$

indicial equation is $r^2 = 0$ so $r = 0$ is a double root. Thus we will obtain only one series of the form

$$y = x^r \sum_{n=0}^{\infty} a_n x^n. \quad \text{The recurrence relation is}$$

$(n+1)^2 a_{n+1} = a_n, \ n = 0,1,2,\ldots$. The coefficients are $a_1 = a_0$, $a_2 = a_1/2^2 = a_0/2^2$, $a_3 = a_2/3^2 = a_0/3^2 \cdot 2^2$, $a_4 = a_3/4^2 = a_0/4^2 \cdot 3^2 \cdot 2^2, \ldots$ and $a_n = a_0/(n!)^2$. Thus one

solution (on setting $a_0 = 1$) is $y = \sum_{n=0}^{\infty} x^n/(n!)^2$.

11. If we make the change of variable $t = x-1$ and let $y = u(t)$, then the Legendre equation transforms to $(t^2 + 2t)u''(t) + 2(t+1)u'(t) - \alpha(\alpha+1)u(t) = 0$. Since $x = 1$ is a regular singular point of the original

equation, we know that $t = 0$ is a regular singular point

of the transformed equation. Substituting $u = \sum_{n=0}^{\infty} a_n t^{n+r}$

in the transformed equation and shifting indices, we obtain

$$\sum_{n=0}^{\infty} (n+r)(n+r-1)a_n t^{n+r} + 2\sum_{n=-1}^{\infty} (n+r+1)(n+r)a_{n+1} t^{n+r}$$

$$+ 2\sum_{n=0}^{\infty} (n+r)a_n t^{n+r} + 2\sum_{n=-1}^{\infty} (n+r+1)a_{n+1} t^{n+r}$$

$$- \alpha(\alpha+1)\sum_{n=0}^{\infty} a_n t^{n+r} = 0, \text{ or}$$

$$[2r(r-1) + 2r]a_0 + \sum_{n=0}^{\infty} \{2(n+r+1)^2 a_{n+1}$$

$$+ [(n+r)(n+r+1) - \alpha(\alpha+1)]a_n\} t^{n+r} = 0.$$

The indicial equation is $2r^2 = 0$ so $r = 0$ is a double root. Thus there will be only one series solution of the

form $y = \sum_{n=0}^{\infty} a_n t^{n+r}$. The recurrence relation is

$2(n+1)^2 a_{n+1} = [\alpha(\alpha+1) - n(n+1)]a_n, n = 0,1,2,\ldots$. We have
$a_1 = [\alpha(\alpha+1)]a_0/2\cdot1^2$, $a_2 = [\alpha(\alpha+1)][\alpha(\alpha+1) - 1\cdot2]a_0/2^2\cdot2^2\cdot1^2$,
$a_3 = [\alpha(\alpha+1)][\alpha(\alpha+1) - 1\cdot2][\alpha(\alpha+1) - 2\cdot3]a_0/2^3\cdot3^2\cdot2^2\cdot1^2,\ldots,$
and $a_n = [\alpha(\alpha+1)][\alpha(\alpha+1)-1\cdot2]\ldots[\alpha(\alpha+1)-(n-1)n]a_0/2^n(n!)^2.$
Reverting to the variable x it follows that one solution of the Legendre equation in powers of x-1 is

$$y_1(x) = \sum_{n=0}^{\infty} [\alpha(\alpha+1)][\alpha(\alpha+1) - 1\cdot2]\ldots$$

$[\alpha(\alpha+1) - (n-1)n](x-1)^n/2^n(n!)^2$ where we have set $a_0 = 1$.

14. The standard form is $y'' + p(x)y' + q(x)y = 0$, with
     $p(x) = 1/x$ and $q(x) = 1$. Thus $x = 0$ is a singular point;
     and since $xp(x) \to 1$ and $x^2 q(x) \to 0$ as $x \to 0$, $x = 0$ is a

regular singular point. Substituting $y = \sum\limits_{n=0}^{\infty} a_n x^{n+r}$ into

$x^2 y'' + xy' + x^2 y = 0$ and shifting indices appropriately, we obtain

$$\sum_{n=0}^{\infty} (n+r)(n+r-1)a_n x^{n+r} + \sum_{n=0}^{\infty} (n+r)a_n x^{n+r} + \sum_{n=2}^{\infty} a_{n-2} x^{n+r} = 0,$$

or

$$[r(r-1)+r]a_0 x^r + [(1+r)r+1+r]a_1 x^{r+1}$$

$$+ \sum_{n=2}^{\infty} [(n+r)^2 a_n + a_{n-2}] x^{n+r} = 0. \quad \text{The indicial equation}$$

is $r^2 = 0$ so $r = 0$ is a double root. It is necessary to take $a_1 = 0$ in order that the coefficient of x^{r+1} be zero. The recurrence relation in $n^2 a_n = -a_{n-2}$, $n = 2,3,\ldots$. Since $a_1 = 0$ it follows that $a_3 = a_5 = a_7 = \ldots = 0$. For the even coefficients we let $n = 2m$, $m = 1,2,\ldots$. Then $a_{2m} = -a_{2m-2}/2^2 m^2$ so $a_2 = -a_0/2^2 \cdot 1^2$, $a_4 = a_0/2^2 \cdot 2^2 \cdot 1^2 \cdot 2^2, \ldots$, and $a_{2m} = (-1)^m a_0/2^{2m}(m!)^2$. Thus one solution of the Bessel

equation of order zero is $J_0(x) = 1 + \sum\limits_{m=1}^{\infty} (-1)^m x^{2m}/2^{2m}(m!)^2$

where we have set $a_0 = 1$. Using the ratio test it can be shown that the series converges for all x. Also note that $J_0(x) \to 1$ as $x \to 0$.

15. In order to determine the form of the integral for x near zero we must study the integrand for x small. Using the above series for J_0, we have

$$\frac{1}{x[J_0(x)]^2} = \frac{1}{x[1 - x^2/2 +\ldots]^2} = \frac{1}{x[1 - x^2 +\ldots]} =$$

$$\frac{1}{x}[1 + x^2 + \ldots] \text{ for x small.} \quad \text{Thus}$$

$$y_2(x) = J_0(x)\int \frac{dx}{x[J_0(x)]^2} = J_0(x)\int [\frac{1}{x} + x + \ldots]dx$$

$$= J_0(x)[\ln x + \frac{x^2}{x} + \ldots],$$

and it is clear that $y_2(x)$ will contain a logarithmic term.

16a. Putting the D.E. in the standard form
$y'' + p(x)y' + q(x)y = 0$ we see that $p(x) = 1/x$ and
$q(x) = (x^2-1)/x^2$. Thus $x = 0$ is a singular point and
since $xp(x) \to 1$ and $x^2q(x) \to -1$ as $x \to 0$, $x = 0$ is a
regular singular point. Substituting $y = \sum_{n=0}^{\infty} a_n x^{n+r}$ into
$x^2 y'' + xy' + (x^2-1)y = 0$, shifting indices appropriately,
and collecting coefficients of common powers of x we
obtain $[r(r-1) + r - 1]a_0 x^r + [(1+r)r + 1 + r - 1]a_1 x^{r+1}$

$$+ \sum_{n=2}^{\infty} \{[(n+r)^2 - 1]a_n + a_{n-2}\}x^{n+r} = 0.$$

The indicial equation is $r^2-1 = 0$ so the roots are $r_1 = 1$
and $r_2 = -1$. For either value of r it is necessary to
take $a_1 = 0$ in order that the coefficient of x^{r+1} be zero.
The recurrence relation is $[(n+r)^2 - 1]a_n = -a_{n-2}$,
$n = 2,3,4\ldots$. For $r = 1$ we have $a_n = -a_{n-2}/[n(n+2)]$,
$n = 2,3,4,\ldots$. Since $a_1 = 0$ it follows that $a_3 = a_5 = a_7$
$= \ldots = 0$. Let $n = 2m$. Then $a_{2m} = -a_{2m-2}/2^2 m(m+1)$, $m =$
$1,2,\ldots$, so $a_2 = -a_0/2^2 \cdot 1 \cdot 2$, $a_4 = -a_2/2^2 \cdot 1 \cdot 2 \cdot 3 =$
$a_0/2^2 \cdot 2^2 \cdot 1 \cdot 2 \cdot 2 \cdot 3,\ldots$, and $a_{2m} = (-1)^m a_0/2^{2m} m!(m+1)!$. Thus one
solution (set $a_0 = 1/2$) of the Bessel equation of order
one is $J_1(x) = (x/2) \sum_{n=0}^{\infty} (-1)^n x^{2n}/(n+1)!n!2^{2n}$. The ratio
test shows that the series converges for all x. Also
note that $J_1(x) \to 0$ as $x \to 0$.

16b. For $r = -1$ the recurrence relation is
$[(n-1)^2 - 1]a_n = -a_{n-2}$, $n = 2,3,\ldots$. Substituting $n = 2$
into the relation yields $[(2-1)^2 - 1]a_2 = 0$ $a_2 = -a_0$.
Hence it is impossible to determine a_2 and consequently
impossible to find a series solution of the form
$$x^{-1} \sum_{n=0}^{\infty} b_n x^n.$$

Secton 5.7, Page 274

1. The D.E. has the form $P(x)y'' + Q(x)y' + R(x)y = 0$ with
    $P(x) = x$, $Q(x) = 2x$, and $R(x) = 6e^x$. From this we find
    $p(x) = Q(x)/P(x) = 2$ and $q(x) = R(x)/P(x) = 6e^x/x$ and
    thus $x = 0$ is a singular point. Since $xp(x) = 2x$ and
    $x^2q(x) = 6xe^x$ are analytic at $x = 0$ we conclude that
    $x = 0$ is a regular singular point. Next, we have
    $xp(x) \to 0 = p_0$ and $x^2q(x) \to 0 = q_0$ as $x \to 0$ and thus the
    indicial equation is $r(r-1) + 0 \cdot r + 0 = r^2 - r = 0$, which
    has the roots $r_1 = 1$ and $r_2 = 0$.

3. The equation has the form $P(x)y'' + Q(x)y' + R(x)y = 0$
    with $P(x) = x(x-1)$, $Q(x) = 6x^2$ and $R(x) = 3$. Since $P(x)$,
    $Q(x)$, and $R(x)$ are polynomials with no common factors and
    $P(0) = 0$ and $P(1) = 0$, we conclude that $x = 0$ and $x = 1$
    are singular points. The first point, $x = 0$, can be shown
    to be a regular singular point using steps similar to
    those to shown in Problem 1. For $x = 1$, we must put the
    D.E. in a form similar to Eq.(1) for this case. To do
    this, divide the D.E. by x and multiply by $(x-1)$ to
    obtain $(x-1)^2y'' + 6x(x-1)y + \dfrac{3}{x}(x-1)y = 0$. Comparing this
    to Eq.(1) we find that $(x-1)p(x) = 6x$ and
    $(x-1)^2q(x) = 3(x-1)/x$ which are both analytic at
    $x = 1$ and hence $x = 1$ is a regular singular point. These
    last two expressions approach $p_0 = 6$ and $q_0 = 0$
    respectively as $x \to 1$, and thus the indicial equation is
    $r(r-1) + 6r + 0 = r(r+5) = 0$.

9. For this D.E., $p(x) = \dfrac{-(1+x)}{x^2(1-x)}$ and $q(x) = \dfrac{2}{x(1-x)}$ and thus
    $x = 0$, -1 are singular points. Since $xp(x)$ is not
    analytic at $x = 0$, $x = 0$ is not a regular singular point.
    Looking at $(x-1)p(x) = \dfrac{1+x}{x^2}$ and $(x-1)^2q(x) = \dfrac{2(1-x)}{x}$ we
    see that $x = 1$ is a regular singular point and that
    $p_0 = 2$ and $q_0 = 0$.

17a. We have $p(x) = \dfrac{\sin x}{x^2}$ and $q(x) = -\dfrac{\cos x}{x^2}$, so that $x = 0$ is
    a singular point. Note that $xp(x) = (\sin x)/x \to 1 = p_0$
    as $x \to 0$ and $x^2q(x) = -\cos x \to -1 = q_0$ as $x \to 0$. In

order to assert that x = 0 is a regular singular point we
must demonstrate that $xp(x)$ and $x^2q(x)$, with $xp(x) = 1$ at
x = 0 and $x^2q(x) = -1$ at x = 0, have convergent power
series (are anlaytic) about x = 0. We know that cosx is
analytic so we need only consider (sinx)/x. Now

$$sinx = \sum_{n=0}^{\infty} x^{2n+1}/(2n+1)! \text{ for } -\infty < x < \infty \text{ so}$$

$$(sinx)/x = \sum_{n=0}^{\infty} x^{2n}/(2n+1)! \text{ and hence is analytic. Thus we}$$

may conclude that x = 0 is a regular singular point.

17b. From part a) it follows that the indicial equation is
$r(r-1) + r - 1 = r^2 - 1 = 0$ and the roots are $r_1 = 1$,
$r_2 = -1$.

17c. To find the first few terms of the solution corresponding
to $r_1 = 1$, assume that
$y = x(a_0 + a_1x + a_2x^2 + ...) = a_0x + a_1x^2 + a_2x^3 + ...$.
Substituting this series for y in the D.E. and expanding
sinx and cosx about x = 0 yields
$x^2(2a_1 + 6a_2x + 12a_3x^2 + 20a_4x^3 + ...) +$
$(x - x^3/3! + x^5/5! + ...)(a_0 + 2a_1x + 3a_2x^2 + 4a_3x^3 + 5a_4x^4 +$
$...) - (1 - x^2/2! + x^4/4! - ...)(a_0x + a_1x^2 + a_2x^3 + a_3x^4 +$
$a_4x^5 + ...) = 0$. Collecting terms, $(2a_1 + 2a_1 - a_1)x^2 +$
$(6a_2 + 3a_2 - a_0/6 - a_2 + a_0/2)x^3 + (12a_3 + 4a_3 - 2a_1/6 - a_3 +$
$a_1/2)x^4 + (20a_4 + 5a_4 - 3a_2/6 + a_0/120 - a_4 + a_2/2 -$
$a_0/24)x^5 + ... = 0$. Simplifying, $3a_1x^2 + (8a_2 + a_0/3)x^3 +$
$(15a_3 + a_1/6)x^4 + (24a_4 - a_0/30)x^5 + ... = 0$. Thus, $a_1 = 0$,
$a_2 = -a_0/4!$, $a_3 = 0$, $a_4 = a_0/6!,...$. Hence
$y_1(x) = x - x^3/4! + x^5/6! + ...$ where we have set $a_0 = 1$.
From Eq. (24) the second solution has the form

$$y_2(x) = ay_1(x)lnx + x^{-1}(1+\sum_{n=1}^{\infty} c_nx^n)$$

$$= ay_1(x)lnx + \frac{1}{x} + c_1 + c_2x + c_3x^2 + c_4x^3 + ..., \text{ so}$$

$y_2' = ay_1'lnx + ay_1x^{-1} - x^{-2} + c_2 + 2c_3x + 3c_4x^2 + ..., \text{ and}$

$y_2'' = ay_1'' \ln x + 2ay_1' x^{-1} - ay_1 x^{-2} + 2x^{-3} + 2c_3 + 3c_4 x + \dots$.
When these are substituted in the given D.E. the terms
including $\ln x$ will appear as
$a[x^2 y_1'' + (\sin x)y_1' - (\cos x)y_1]$, which is zero since y_1 is
a solution. For the remainder of the terms, use
$y_1 = x - x^3/24 + x^5/720$ and the $\cos x$ and $\sin x$ x series as
shown earlier to obtain
$-c_1 + (2/3+2a)x + (3c_3+c_1/2)x^2 + (4/45+c_2/3+8c_4)x^3 +\dots= 0$.
These yield $c_1 = 0$, $a = -1/3$, $c_3 = 0$, and
$c_4 = -c_2/24 - 1/90$. We may take $c_2 = 0$, since this term
will simply generate $y_1(x)$ over again. Thus
$y_2(x) = -\dfrac{1}{3}y_1(x)\ln x + x^{-1} - \dfrac{1}{90}x^3$. If a computer algebra
system is used, then additional terms in each series may
be obtained without much additional effort. The next
terms, in each case, are shown here:
$y_1(x) = x - \dfrac{x^3}{24} + \dfrac{x^5}{720} - \dfrac{43x^7}{1451520} + \dots$ and
$y_2(x) = -\dfrac{1}{3}y_1(x)\ln x + \dfrac{1}{x}[1 - \dfrac{x^4}{90} + \dfrac{41x^6}{120960} - \dots]$.

18. We first write the D.E. in the standard form as given for
Theorem 5.7.1 except that we are expanding in powers of
$(x-1)$ rather than powers of x:
$(x-1)^2 y'' + (x-1)[(x-1)/2\ln x]y' + [(x-1)^2/\ln x]y = 0$. since
$\ln 1 = 0$, $x = 1$ is a singular point. To show it is a
regular singular point of this D.E. we must show that
$(x-1)/\ln x$ is analytic at $x = 1$; it will then follow that
$(x-1)^2/\ln x = (x-1)[(x-1)/\ln x]$ is also analytic at
$x = 1$. If we expand $\ln x$ in a Taylor series about $x = 1$
we find that $\ln x = (x-1) - \dfrac{1}{2}(x-1)^2 + \dfrac{1}{3}(x-1)^3 - \dots$.

Thus $(x-1)/\ln x = [1 - \dfrac{1}{2}(x-1) + \dfrac{1}{3}(x-1)^2 - \dots]^{-1} =$

$1 + \dfrac{1}{2}(x-1) + \dots$ has a power series expansion about

$x = 1$, and hence is analytic. We can use the above
result to obtain the indicial equation at $x = 1$. We have

$(x-1)^2 y'' + (x-1)[\dfrac{1}{2} + \dfrac{1}{4}(x-1) + \dots]y' + [(x-1) +$

$\dfrac{1}{2}(x-1)^2 + \dots]y = 0$. Thus $p_0 = 1/2$, $q_0 = 0$ and the
indicial equation is $r(r-1) + r/2 = 0$. Hence $r = 1/2$ and

$r = 0$. In order to find the first three non-zero terms in a series solution corresponding to $r = 1/2$, it is better to keep the differential equation in its original form and to substitute the above power series for $\ln x$:

$$[(x-1) - \frac{1}{2}(x-1)^2 + \frac{1}{3}(x-1)^3 - \frac{1}{4}(x-1)^4 + \ldots]y'' + \frac{1}{2}y' + y = 0.$$

Next we substitute $y = a_0(x-1)^{1/2} + a_1(x-1)^{3/2} + a_2(x-1)^{5/2} + \ldots$ and collect coefficients of like powers of $(x-1)$ which are then set equal to zero. This requires some algebra before we find that $6a_1/4 + 9a_0/8 = 0$ and $5a_2 + 5a_1/8 - a_0/12 = 0$. These equations yield $a_1 = -3a_0/4$ and $a_2 = 53a_0/480$. With $a_0 = 1$ we obtain the solution

$$y_1(x) = (x-1)^{1/2} - \frac{3}{4}(x-1)^{3/2} + \frac{53}{480}(x-1)^{5/2} + \ldots .$$ Since the radius of convergence of the series for $\ln x$ is 1, we would expect $\rho = 1$.

20a. If we write the D.E. in the standard form as given in Theorem 5.7.1 we obtain $x^2 y'' + x[\alpha/x]y' + [\beta/x]y = 0$ where $xp(x) = \alpha/x$ and $x^2 q(x) = \beta/x$. Neither of these terms are analytic at $x = 0$ so $x = 0$ is an irregular singular point.

20b. Substituting $y = x^r \sum\limits_{n=0}^{\infty} a_n x^n$ in $x^3 y'' + \alpha xy' + \beta y = 0$ gives

$$\sum_{n=0}^{\infty} (n+r)(n+r-1)a_n x^{n+r+1} + \alpha \sum_{n=0}^{\infty} (n+r)a_n x^{n+r} + \beta \sum_{n=0}^{\infty} a_n x^{n+r} = 0.$$

Shifting the index in the first series and collecting coefficients of common powers of x we obtain $(\alpha r + \beta)a_0 x^r$

$$+ \sum_{n=1}^{\infty} (n+r-1)(n+r-2)a_{n-1} + [\alpha(n+r) + \beta]a_n x^{n+r} = 0.$$ Thus the indicial equation is $\alpha r + \beta = 0$ with the single root $r = -\beta/\alpha$.

20c. From part b, the recurrence relation is

$$a_n = \frac{(n+r-1)(n+r-2)a_{n-1}}{\alpha(n+r) + \beta}, \quad n = 1, 2, \ldots$$

$$= \frac{(n - \frac{\beta}{\alpha} - 1)(n - \frac{\beta}{\alpha} - 2)a_{n-1}}{\alpha n}.$$

For $\dfrac{\beta}{\alpha} = -1$, then, $a_n = \dfrac{n(n-1)a_{n-1}}{\alpha n}$, which is zero for
$n = 1$ and thus $y(x) = x$ is the solution. Similarly for
$\dfrac{\beta}{\alpha} = 0$, $a_n = \dfrac{(n-1)(n-2)}{\alpha n}$ and again for $n = 1$ $a_1 = 0$ and
$y(x) = 1$ is the solution. Continuing in this fashion, we
see that the series solution will terminate for β/α any
positive integer as well as 0 and -1. For other values
of β/α, we have $\dfrac{a_n}{a_{n-1}} = \dfrac{(n-\frac{\beta}{2}-1)(n-\frac{\beta}{\alpha}-2)}{\alpha n}$, which approaches
∞ as $n \to \infty$ and thus the ratio test yields a zero radius
of convergence.

21b. Substituting $y = \displaystyle\sum_{n=0}^{\infty} a_n x^{n+r}$ in the D.E. in standard form

(gives

$$\sum_{n=0}^{\infty} (n+r)(n+r-1)a_n x^{n+r} + \alpha \sum_{n=0}^{\infty} (n+r)a_n x^{n+r+1-s}$$

$$+ \beta \sum_{n=0}^{\infty} a_n x^{n+r+2-t} = 0.$$

If $s = 2$ and $t = 2$ the first term in each of the three
series is $r(r-1)a_0 x^r$, $\alpha r a_0 x^{r-1}$, and $\beta a_0 x^r$, respectively.
Thus we must have $\alpha r a_0 = 0$ which requires $r = 0$. Hence
there is at most one solution of the assumed form.

21d. In order for the indicial equation to be quadratic in r
it is necessary that the first term in the first series
contribute to the indicial equation. This means that the
first term in the second and the third series cannot
appear before the first term of the first series. The
first terms are $r(r-1)a_0 x^r$, $\alpha r a_0 x^{r+1-s}$, and $\beta a_0 x^{r+2-t}$,
respectively. Thus if $s \le 1$ and $t \le 2$ the quadratic term
will appear in the indicial equation.

Section 5.8, Page 285

1. It is clear that $x = 0$ is a singular point. The D.E. is
in the standard form given in Theorem 5.7.1 with
$xp(x) = 2$ and $x^2 q(x) = x$. Both are analytic at $x = 0$, so

x = 0 is a regular singular point.   Substituting

$$y = \sum_{n=0}^{\infty} a_n x^{n+r} \text{ in the D.E., shifting indices}$$

appropriately, and collecting coefficients of like powers
of x yields

$$[r(r-1) + 2r]a_0 x^r + \sum_{n=1}^{\infty} [(r+n)(r+n+1)a_n + a_{n-1}]x^{r+n} = 0.$$

The indicial equation is $F(r) = r(r+1) = 0$ with roots
$r_1 = 0$, $r_2 = -1$.   Treating a_n as a function of r, we see
that $a_n(r) = -a_{n-1}(r)/F(r+n)$, $n = 1,2,...$ if $F(r+n) \neq 0$.
Thus $a_1(r) = -a_0/F(r+1)$, $a_2(r) = a_0/F(r+1)F(r+2),...,$ and
$a_n(r) = (-1)^n a_0/F(r+1)F(r+2)...F(r+n)$, provided $F(r+n) \neq$
0 for $n = 1,2,...$.   For the case $r_1 = 0$, we have
$a_n(0) = (-1)^n a_0/F(1)F(2) \ ... \ F(n) = (-1)^n a_0/n!(n+1)!$ so

$$\text{one solution is } y_1(x) = \sum_{n=0}^{\infty} (-1)^n x^n/n!(n+1)! \text{ where we have}$$

set $a_0 = 1$.

If we try to use the above recurrence relation for
the case $r_2 = -1$ we find that $a_n(-1) = -a_{n-1}/n(n-1)$,
which is undefined for $n = 1$.   Thus we must follow the
procedure described at the end of Section 5.7 to
calculate a second solution of the form given in Eq.(24).
Specifically, we use Eqs.(19) and (20) of that section to
calculate a and $c_n(r_2)$ where $r_2 = -1$.   Since
$r_1 - r_2 = 1 = N$, we have $a_N(r) = a_1(r) = -1/F(r+1)$.Hence

$$a = \lim_{r \to -1} [(r+1)(-1)/F(r+1)] = \lim_{r \to -1} [-(r+1)/(r+1)(r+2)] = -1.$$

Next

$$c_n(-1) = \frac{d}{dr}[(r+1)a_n(r)]\Big|_{r=-1} = (-1)^n \frac{d}{dr}[\frac{(r+1)}{F(r+1) \ ... \ F(r+n)}]\Big|_{r=-1},$$

where we have set $a_0 = 1$.   Observe that $(r+1)/F(r+1) \ ...$
$F(r+n) = 1/[(r+2)^2(r+3)^2...(r+n)^2(r+n+1)] = 1/G_n(r)$.
Hence $c_n(-1) = (-1)^{n+1} G_n'(-1)/G_n^2(-1)$.   Notice that
$G_n(-1) = 1^2 \cdot 2^2 \cdot 3^2 ... (n-1)^2 n = (n-1)!n!$ and
$G_n'(-1)/G_n(-1) = 2[1/1 + 1/2 + 1/3 +...+ 1/(n-1)] + 1/n =$
$H_n + H_{n-1}$.   Thus $c_n(-1) = (-1)^{n+1}(H_n + H_{n-1})/(n-1)!n!$.
From Eq.(24) of Section 5.7 we obtain the second solution

$$y_2(x) = -y_1(x)\ln x + x^{-1}\left[1 - \sum_{n=1}^{\infty}(-1)^n(H_n + H_{n-1})x^n/n!\,(n-1)!\right].$$

2.   It is clear that $x = 0$ is a singular point. The D.E. is in the standard form given in Theorem 5.7.1 with $xp(x) = 3$ and $x^2q(x) = 1+x$. Both are analytic at $x = 0$, so $x = 0$ is a regular singular point. Substituting

$$y = \sum_{n=0}^{\infty} a_n x^{n+r}$$ in the D.E., shifting indices

appropriately, and collecting coefficients of like powers of x yields

$$[r(r-1) + 3r + 1]a_0 x^r + \sum_{n=1}^{\infty}\{[(r+n)(r+n+2) + 1]a_n$$

$$+ a_{n-1}\}\, x^{n+r} = 0.$$

The indicial equation is $F(r) = r^2 + 2r + 1 = (r+1)^2 = 0$ with the double root $r_1 = r_2 = -1$. Treating a_n as a function of r, we see that $a_n(r) = -a_{n-1}(r)/F(r+n)$, $n = 1,2,\ldots$. Thus $a_1(r) = -a_0/F(r+1)$, $a_2(r) = a_0/F(r+1)F(r+2),\ldots$, and $a_n(r) = (-1)^n a_0/F(r+1)F(r+2)\ldots F(r+n)$. Setting $r = -1$ we find that $a_n(-1) = (-1)^n a_0/(n!)^2$, $n = 1,2,\ldots$. Hence one solution is $y_1(x) = x^{-1}\sum_{n=0}^{\infty}(-1)^n x^n/(n!)^2$ where we have set $a_0 = 1$. To find a second solution we follow the procedure described in Section 5.7 for the case when the roots of the indicial equation are equal. Specifically, the second solution will have the form given in Eq.(17) of that section. We must calculate $a_n'(-1)$. If we let $G_n(r) = F(r+1)\ldots F(r+n) = (r+2)^2(r+3)^2\ldots(r+n+1)^2$ and take $a_0 = 1$, then $a_n'(-1) = (-1)^n[1/G_n(r)]'$ evaluated $r = -1$. Hence $a_n'(-1) = (-1)^{n+1}G_n'(-1)/G_n^2(-1)$. But $G_n(-1) = (n!)^2$ and $G_n'(-1)/G_n(-1) = 2[1/1 + 1/2 + 1/3 + \ldots + 1/n] = 2H_n$. Thus a second solution is

$$y_2(x) = y_1(x)\ln x - 2x^{-1}\sum_{n=1}^{\infty}(-1)^n H_n x^n/(n!)^2.$$

3.   The roots of the indicial equation are r_1 and $r_2 = 0$ and
     thus the analysis is similar to that for Problem 1b.

4.   The roots of the indicial equation are $r_1 = -1$ and
     $r_2 = -2$ and thus the analysis is similar to that for
     Problem 1.

5.   Since $x = 0$ is a regular singular point, substitute

$$y = \sum_{n=0}^{\infty} a_n x^{n+r}$$ in the D.E., shift indices appropriately,

and collect coefficients of like powers of x to obtain
$[r^2 - 9/4]a_0 x^r + [(r+1)^2 - 9/4]a_1 x^{r+1}$

$$+ \sum_{n=2}^{\infty} \{[(r+n)^2 - 9/4]a_n + a_{n-2}\} \, x^{n+r} = 0.$$

The indicial equation is $F(r) = r^2 - 9/4 = 0$ with roots
$r_1 = 3/2$, $r_2 = -3/2$. Treating a_n as a function of r we
see that $a_n(r) = -a_{n-2}(r)/F(r+n)$, $n = 2,3,..$ if
$F(r+n) \neq 0$. For the case $r_1 = 3/2$, $F(r_1+1)$, which is the
coefficient of x^{r_1+1} is $\neq 0$ so we must set $a_1 = 0$. It
follows that
$a_3 = a_5 = ... = 0$. For the even coefficients, set $n = 2m$
so $a_{2m}(3/2) = -a_{2m-2}(3/2)/F(3/2 + 2m)$, $m = 1,2...$. Thus
$a_2(3/2) = - a_0/2^2 \cdot 1(1 + 3/2)$,

$a_4(3/2) = a_0/2^4 \cdot 2!(1 + 3/2)(2 + 3/2),...$, and

$a_{2m}(3/2) = (-1)^m/2^{2m}m! \cdot (1 + 3/2)...(m + 3/2)$. Hence one
solution is

$$y_1(x) = x^{3/2}[1 + \sum_{m=1}^{\infty} \frac{(-1)^m}{m!(1 + 3/2)(2 + 3/2)...(m + 3/2)} (\frac{x}{2})^{2m}],$$

where we have set $a_0 = 1$. For this problem, the roots r_1
and r_2 of the indicial equation differ by an integer:
$r_1 - r_2 = 3/2 - (-3/2) = 3$. Hence we can anticipate that
there may be difficulty in calculating a second solution
corresponding to $r = r_2$. This difficulty will occur in
calculating $a_3(r) = - a_1(r)/F(r+3)$ because when
$r = r_2 = -3/2$ we have $F(r_2+3) = F(r_1) = 0$. However, in
this problem we are fortunate because $a_1 = 0$ and it will
not be necessary to use the theory described at the end
of Section 5.7. Notice for $r = r_2 = -3/2$ that the

coefficient of s^{r_2+1} is $[(r_2+1)^2 - 9/4]a_1$, which does not vanish unless $a_1 = 0$. Thus the recurrence relation for the odd coefficients yields $a_5 = -a_3/F(7/2)$, $a_7 = -a_5/F(11/2) = a_3/F(11/2)F(7/2)$ and so forth. Substituting these terms into the assumed form we see that a multiple of $y_1(x)$ has been obtained and thus we may take $a_3 = 0$ without loss of generality. Hence $a_3 = a_5 = a_7 = \ldots = 0$. The even coefficients are given by $a_{2m}(-3/2) = -a_{2m-2}(-3/2)/F(2m - 3/2)$, $m = 1,2\ldots$.

Thus $a_2(-3/2) = -a_0/2^2 \cdot 1 \cdot (1 - 3/2)$,

$a_4(-3/2) = a_0/2^4 \cdot 2!(1 - 3/2)(2 - 3/2),\ldots$, and

$a_{2m}(-3/2) = (-1)^m a_0/2^{2m} m!(1 - 3/2)(2 - 3/2) \ldots (m - 3/2)$.

Thus a second solution is

$$y_2(x) = x^{-3/2}[1 + \sum_{m=1}^{\infty} \frac{(-1)^m}{m!(1 - 3/2)(2 - 3/2) \ldots (m - 3/2)} (\frac{x}{2})^{2m}].$$

7.   Apply the ratio test:

$$\lim_{m \to \infty} \frac{|(-1)^{m+1} x^{2m+2}/2^{2m+2}[(m+1)!]^2|}{|(-1)^m x^{2m}/2^{2m}(m!)^2|} = |x^2| \lim_{m \to \infty} \frac{1}{2^2(m+1)^2} = 0$$

for every x. Thus the series for $J_0(x)$ converges absolutely for all x.

12.   If $\xi = \alpha x^\beta$, then $dy/dx = \frac{1}{2}x^{-1/2}f + x^{1/2}f'\alpha\beta x^{\beta-1}$ where f' denotes $df/d\xi$. Find d^2y/dx^2 in a similar fashion and use algebra to show that f satisfies the D.E.

$\xi^2 f'' + \xi f' + [\xi^2 - \upsilon^2]f = 0$.

13.   To compare $y'' - xy = 0$ with the D.E. of Problem 12, we must multiply by x^2 to get $x^2 y'' - x^3 y = 0$. Thus $2\beta = 3$, $\alpha^2\beta^2 = -1$ and $1/4 - \upsilon^2\beta^2 = 0$. Hence $\beta = 3/2$, $\alpha = 2i/3$ and $\upsilon = 1/3$ which yields the desired result.

14.   First we verify that $J_0(\lambda_j x)$ satisfies the D.E. We know that $J_0(t)$ is a solution of the Bessel equation of order zero:

$t^2 J_0''(t) + t J_0'(t) + t^2 J_0(t) = 0$ or

$J_0''(t) + t^{-1} J_0'(t) + J_0(t) = 0$.

Let $t = \lambda_j x$. Then

$$\frac{d}{dx} J_0(\lambda_j x) = \frac{d}{dt} J_0(t) \frac{dt}{dx} = \lambda_j J_0'(t)$$

$$\frac{d^2}{dx^2} J_0(\lambda_j x) = \lambda_j \frac{d}{dt}[J_0'(t)] \frac{dt}{dx} = \lambda_j^2 J_0''(t).$$

Substituting $y = J_0(\lambda_j x)$ in the given D.E. and making use of these results, we have

$$\lambda_j^2 J_0''(t) + (\lambda_j/t) \, \lambda_j J_0'(t) + \lambda_j^2 J_0(t) =$$

$$\lambda_j^2[J_0''(t) + t^{-1} J_0'(t) + J_0(t)] = 0.$$

Thus $y = J_0(\lambda_j x)$ is a solution of the given D.E. For the second part of the problem we follow the hint. First, rewrite the D.E. by multiplying by x to yield $xy'' + y' + \lambda_j^2 xy = 0$, which can be written as $(xy')' = -\lambda_j^2 xy$. Now let $y_i(x) = J_0(\lambda_i x)$ and $y_j(x) = J_0(\lambda_j x)$ and we have, respectively:

$$(xy_i')' = -\lambda_i^2 xy_i$$

$$(xy_j')' = -\lambda_j^2 xy_j.$$

Now multiply the first equation by y_j, the second by y_i, integrate each from 0 to 1, and subtract the second from the first:

$$\int_0^1 [y_j(xy_i')' - y_i(xy_j')'] dx = -(\lambda_i^2 - \lambda_j^2) \int_0^1 xy_i y_j dx.$$

If we integrate each term on the left side once by parts and note that $y_i = y_j = 0$ at $x = 0$ and $x = 1$, we find that the left side of this equation is identically zero. Hence the right side is identically zero and for $\lambda_i \neq \lambda_j$ this gives the desired result.

CHAPTER 6

Section 6.1, Page 294

1. The graph of $f(t)$ is shown.
    Since the function is
    continuous on each interval,
    but has a jump discontinuity
    at $t = 1$, $f(t)$ is piecewise
    continuous.

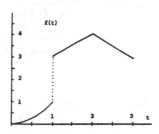

5b. Since t^2 is continuous for $0 \leq t \leq A$ for any positive A
    and since $t^2 \leq e^{at}$ for any $a > 0$ and for t sufficiently
    large, it follows from Theorem 6.1.2 that $\mathcal{L}\{t^2\}$ exists
    for $s > 0$. $\mathcal{L}\{t^2\} = \int_0^\infty e^{-st}t^2 dt = \lim_{M \to \infty} \int_0^M e^{-st}t^2 dt$

$$= \lim_{M \to \infty} [\frac{-t^2}{s}e^{-st}|_0^M + \frac{2}{s}\int_0^M e^{-st}t dt]$$

$$= \frac{2}{s}\lim_{M \to \infty} [-\frac{1}{s}te^{-st}|_0^M + \frac{1}{s}\int_0^M e^{-st}dt]$$

$$= \frac{2}{s^2}\lim_{M \to \infty} - \frac{1}{s}e^{-st}|_0^M = \frac{2}{s^3}.$$

6. That $f(t) = \cos at$ satisfies the hypotheses of Theorem
    6.1.2 can be verified by recalling that $|\cos at| \leq 1$ for
    all t. To determine $\mathcal{L}\{\cos at\} = \int_0^\infty e^{-st} \cos at dt$ we
    must integrate by parts twice to get $\int_0^\infty e^{-st} \cos at dt =$
    $\lim_{M \to \infty} [(-s^{-1}e^{-st} \cos at + as^{-2} e^{-st} \sin at)|_0^M$
    $- (a^2/s^2) \int_0^M e^{-st} \cos at dt]$. Evaluating the first two
    terms, letting $M \to \infty$, and adding the third term to both
    sides, we obtain $[1 + a^2/s^2]\int_0^\infty e^{-st} \cos at dt = 1/s$, $s > 0$.
    Division by $[1 + a^2/s^2]$ and simplification yields the
    desired solution.

9. From the definition for $\cosh bt$ we have
    $\mathcal{L}\{e^{at}\cosh bt\} = \mathcal{L}\{\frac{1}{2}[e^{(a+b)t} + e^{(a-b)t}]\}$. Using the linearity
    property of $\mathcal{L}$, Eq.(5), the right side becomes
    $\frac{1}{2}\mathcal{L}\{e^{(a+b)t}\} + \frac{1}{2}\mathcal{L}\{e^{(a-b)t}\}$ which can be evaluated using the

result of Example 5 and thus

$$\mathcal{L}\{e^{at}\cosh bt\} = \frac{1/2}{s-(a+b)} + \frac{1/2}{s-(a-b)}$$

$$= \frac{s-a}{(s-a)^2 - b^2}, \quad \text{for } s-a > |b|.$$

13. We write $\sin at = (e^{iat} - e^{-iat})/2i$, then the linearity of
     the Laplace transform operator allows us to write
     $\mathcal{L}\{e^{at}\sin bt\} = (1/2i)\mathcal{L}\{e^{(a+ib)t}\} - (1/2i)\mathcal{L}\{e^{(a-ib)t}\}$. Each of
     these two terms can be evaluated by using the result of
     Example 5, where we now have to require s to be greater
     than the real part of the complex numbers $a \pm ib$ in order
     for the integrals to converge. Complex algebra then
     gives the desired result. An alternate method of
     evaluation would be to use integration on the integral
     appearing in the definition of $\mathcal{L}\{e^{at}\sin bt\}$, but that
     method requires integration by parts twice.

16. As in Problem 13, $\mathcal{L}\{t\sin at\} = (1/2i)\mathcal{L}\{te^{iat}\} -$
     $(1/2i)\mathcal{L}\{te^{-iat}\}$. Using the result of Problem 15 we obtain
     $\mathcal{L}\{t\sin at\} = (1/2i)[(s-b)^{-2} - (s+b)^{-2}]$ where $b = ia$ and
     $s > 0$. Hence $\mathcal{L}\{t\sin at\} = 2as/(s^2+a^2)^2$, $s > 0$.

19. Use the approach shown in Problem 16 with the result of
     Problem 18, for $n = 2$. A computer algebra system may
     also be used.

21. The integral $\int_0^A (t^2 + 1)^{-1}dt$ can be evaluated in terms of
     the arctan function and then Eq. (3) can be used. To
     illustrate Theorem 6.1.1, however, consider that
     $\frac{1}{t^2+1} < \frac{1}{t^2}$ for $t \geq 1$ and, from Example 3, $\int_1^\infty t^{-2}dt$
     converges and hence $\int_1^\infty (t^2 + 1)^{-1}dt$ also converges.
     $\int_0^1 (t^2 + 1)^{-1}dt$ is finite and hence does not affect the
     convergence of $\int_0^\infty (t^2 + 1)^{-1}dt$ at infinity.

25. If we let $u = f$ and $dv = e^{-st}dt$ then $F(s) = \int_0^\infty e^{-st}f(t)dt$
     $= \lim_{M \to \infty} -\frac{1}{s}e^{-st}f(t)|_0^M + \frac{1}{s}\int_0^\infty e^{-st}f'(t)dt$. Now use an
     argument similar to that given to establish Theorem 6.1.2.

27a. Make a transformation of variables with x = st and
     dx = sdt. Then use the definition of $\Gamma(P+1)$ from
     Problem 26.

27d. Use the definition of $\mathcal{L}\{t^{1/2}\}$ and integrate by parts once
     to get $\mathcal{L}\{t^{1/2}\} = (1/2s)\mathcal{L}\{t^{-1/2}\}$. The result follows from
     part (c).

Section 6.2, Page 303

Problems 1 through 10 are solved by using partial fractions
and algebra to manipulate the given function into a form
matching one of the functions appearing in the middle column
of Table 6.2.1.

2.   We have $\dfrac{4}{(s-1)^3} = 2\dfrac{2!}{(s-1)^{2+1}}$ and thus the inverse Laplace
     transform is $2t^2e^t$, using line 11.

4.   We have $\dfrac{3s}{s^2-s-6} = \dfrac{3s}{(s-3)(s+2)} = \dfrac{9/5}{s-3} + \dfrac{6/5}{s+2}$ using partial
     fractions. Thus $(9/5)e^{3t} + (6/5)e^{-2t}$ is the inverse
     transform, from line 2.

7.   We have $\dfrac{2s+1}{s^2-2s+2} = \dfrac{2s+1}{(s-1)^2+1} = \dfrac{2(s-1)}{(s-1)^2+1} + \dfrac{3}{(s-1)^2+1}$, where
     we first used the concept of completing the square (in
     the denominator) and then added and subtracted
     appropriately to put the numerator in the desired form.
     Lines 9 and 10 may now be used to find the desired
     result.

In each of the Problems 11 through 23 it is assumed that the
I.V.P. has a solution $y = \phi(t)$ which, with its first two
derivatives, satisfies the conditions of the Corollary to
Theorem 6.2.1.

11. Take the Laplace transform of the D.E. to get
     $s^2Y(s) - sy(0) - y'(0) - [sY(s) - y(0)] - 6Y(s) = 0$.
     Using the I.C. and solving for Y(s) we obtain
     $Y(s) = \dfrac{s-2}{s^2-s-6}$. Following the pattern of Eq.(12) we have
     $\dfrac{s-2}{s^2-s-6} = \dfrac{a}{s+2} + \dfrac{b}{s-3} = \dfrac{a(s-3)+b(s+2)}{(s+2)(s-3)}$. Equating like

powers in the numerators we find a+b = 1 and
-3a + 2b = -2. Thus a = 4/5 and b = 1/5 and
$Y(s) = \dfrac{4+5}{s+2} + \dfrac{1/5}{s-3}$, which yields the desired solution
using Table 6.2.1.

14. Taking the Laplace transform we have $s^2Y(s) - sy(0) -$
y'(0) - 4[sY(s)-y(0)] + 4Y(s) = 0. Using the I.C. and
solving for Y(s) we find $Y(s) = \dfrac{s-3}{s^2-4s+4}$. Since the
denominator is a perfect square, the partial fraction
form is $\dfrac{s-3}{s^2-4s+4} = \dfrac{a}{(s-2)^2} + \dfrac{b}{s-2}$. Solving for a and b,
as shown in examples of this section or in Problem 11, we
find a = -1 and b = 1. Thus $Y(s) = \dfrac{1}{s-2} - \dfrac{1}{(s-2)^2}$, from
which we find $y(t) = e^{2t} - te^{2t}$.

15. Note that $Y(s) = \dfrac{2s-4}{s^2-2s-2} = \dfrac{2s-4}{(s-1)^2-3} = \dfrac{2(s-1)}{(s-1)^2-3} - \dfrac{2}{(s-1)^2-3}$.
Three formulas in Table 6.2.1 are now needed: F(s-c) in
line 14 in conjunction with the ones for coshat and
sinhat, lines 7 and 8.

17. The Laplace transform of the D.E. is
$s^4Y(s) - s^3y(0) - s^2y'(0) - sy''(0) - y'''(0) - 4[s^3Y(s)-s^2y(0)$
$-sy'(0) - y''(0)] + 6[s^2Y(s) - sy(0) - y'(0)] - 4[sY(s) - y(0)]$
+Y(s) = 0. Using the I.C. and solving for Y(s) we find
$Y(s) = \dfrac{s^2 - 4s + 7}{s^4-4s^3+6s^2-4s+1}$. The correct partial fraction
form for this is $\dfrac{a}{(s-1)^4} + \dfrac{b}{(s-1)^3} + \dfrac{c}{(s-1)^2} + \dfrac{d}{s-1}$.
Setting this equal to Y(s) above and equating the
numerators we have $s^2-4s+7 = a + b(s-1) + c(s-1)^2 +$
$d(s-1)^3$. Solving for a,b,c, and d and use of Table 6.2.1
yields the desired solution.

20. The Laplace transform of the D.E. is
$s^2Y(s) - sy(0) - y'(0) + \omega^2Y(s) = s/(s^2+4)$. Applying the
I.C. and solving for Y(s) we get $Y(s) = s/[(s^2+4)(s^2+\omega^2)]$
$+ s/(s^2+\omega^2)$. Decomposing the first term by partial
fractions we have

$$Y(s) = \frac{s}{(\omega^2-4)(s^2+4)} - \frac{s}{(\omega^2-4)(s^2+\omega^2)} + \frac{s}{s^2+\omega^2}$$

$$= (\omega^2-4)^{-1}[\frac{(\omega^2-5)s}{s^2+\omega^2} + \frac{s}{s^2+4}].$$

Then, using Table 6.1.2, we have

$$y = (\omega^2-4)^{-1}[(\omega^2-5)\cos\omega t + \cos 2t].$$

22. Solving for $Y(s)$ we find $Y(s) = 1/[(s-1)^2 + 1] + 1/(s+1)[(s-1)^2 + 1]$. Using partial fractions on the second term we obtain

$$Y(s) = 1/[(s-1)^2 + 1] + \{1/(s+1) - (s-3)/[(s-1)^2 + 1]\}/5$$

$$= (1/5)\{(s+1)^{-1} - (s-1)[(s-1)^2 + 1]^{-1} + 7[(s-1)^2 + 1]^{-1}\}.$$

Hence, $y = (1/5)(e^{-t} - e^t\cos t + 7e^t\sin t)$.

24. Under the standard assumptions, the Lapace transform of the left side of the D.E. is $s^2Y(s) - sy(0) - y'(0) + 4Y(s)$. To transform the right side we must revert to the definition of the Laplace trasnform to determine $\int_0^\infty e^{-st}f(t)dt$. Since $f(t)$ is piecewise continuous we are able to calculate $\mathcal{L}\{f(t)\}$ by

$$\int_0^\infty e^{-st}f(t)dt = \int_0^\pi e^{-st}dt + \lim_{M \to \infty}\int_\pi^M (e^{-st})(0)dt$$

$$= \int_0^\pi e^{-st}dt = (1 - e^{-\pi s})/s.$$

Hence, the Laplace transform $Y(s)$ of the solution is given by $Y(s) = s/(s^2+4) + (1 - e^{-\pi s})/s(s^2+4)$.

27b. The Taylor series for f about $t = 0$ is

$$f(t) = \sum_{n=0}^\infty (-1)^n t^{2n}/(2n+1)!, \text{ which is obtained from}$$

part(a) by dividing each term of the sine series by t. Also, f is continuous for $t > 0$ since $\lim_{t \to 0+}(\sin t)/t = 1$.

Assuming that we can compute the Laplace transform of f term by term, we obtain $\mathcal{L}\{f(t)\} = \mathcal{L}\{\sum_{n=0}^\infty (-1)^n t^{2n}/(2n+1)!\}$

$$= \sum_{n=0}^\infty [(-1)^n/(2n+1)!\mathcal{L}\{t^{2n}\}$$

$$= \sum_{n=0}^\infty [(-1)^n(2n)!/(2n+1)!]s^{-(2n+1)}$$

$$= \sum_{n=0}^{\infty} [(-1)^n/(2n+1)]s^{-(2n+1)},$$ which converges for s > 1.

The Taylor series for arctan x is given by

$$\sum_{n=0}^{\infty} (-1)^n \, x^{2n+1}/(2n+1).$$ Comparing $\mathcal{L}\{f(t)\}$ with the Taylor

series for arctanx, we conclude that
$\mathcal{L}\{f(t)\} = \arctan(1/s)$, s > 1.

30. Setting n = 2 in Problem 28b, we have

$$\mathcal{L}\{t^2 \sin bt\} = \frac{d^2}{ds^2}[b/(s^2+b^2)] = \frac{d}{ds}[-2bs/(s^2+b^2)^2] =$$

$$-2b/(s^2+b^2)^2 + 8bs^2/(s^2+b^2)^3 = 2b(3s^2-b^2)/(s^2+b^2)^3.$$

32. Using the result of Problem 28a. we have

$$\mathcal{L}\{te^{at}\} = -\frac{d}{ds}(s-a)^{-1} = (s-a)^{-2}$$

$$\mathcal{L}\{t^2 e^{at}\} = -\frac{d}{ds}(s-a)^{-2} = 2(s-a)^{-3}.$$

$$\mathcal{L}\{t^3 e^{at}\} = -\frac{d}{ds}2(s-a)^{-3} = 3!(s-a)^{-4}.$$ Continuing in this

fashion, or using induction, we obtain the desired
result.

36a. Taking the Laplace transform of the D.E. we obtain
$\mathcal{L}\{y''\} - \mathcal{L}\{ty\} = \mathcal{L}\{y''\} + \mathcal{L}\{-ty\}$
$$= s^2 Y(s) - sy(0) - y'(0) + Y'(s) = 0.$$
Hence, Y satisfies $Y' + s^2 Y = s$.

38a. Follow the hint and apply L'Hopital's rule, recalling
that Q(s) has distinct zeros.

Section 6.3, Page 310

2.   From the definition of $u_c(t)$
     we have:
     $g(t) = (t-3)u_2(t) - (t-2)u_3(t)$

$$= \begin{cases} 0 - 0 = 0, & 0 \le t < 2 \\ (t-3) - 0 = t-3, & 2 \le t < 3. \\ (t-3) - (t-2) = -1, & 3 \le t \end{cases}$$

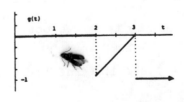

4.   As indicated in the discussion
     following Example 1, the unit
     step function can be used to
     translate a given function f,
     with domain t≥0, a distance c
     to the right by the
     multiplication $u_c(t)f(t-c)$.

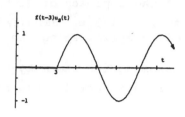

     Hence the required graph of
     $y = f(t-3)u_3(t)$ for $f(t) = \sin t$ is shown.

8.   In order to use Theorem 6.3.1 we must write $f(t)$ in terms
     of $u_c(t)$. Since $t^2 - 2t + 2 = (t-1)^2 + 1$ (by completing
     the square), we can thus write $f(t) = u_1(t)g(t-1)$, where
     $g(t) = t^2+1$. Now applying Theorem 6.3.1 we have
     $\mathcal{L}\{f(t)\} = \mathcal{L}\{u_1(t)g(t-1)\} = e^{-s}\mathcal{L}\{g(t)\} = e^{-s}(2/s^3 + 1/s)$.

14. Use partial fractions to write
     $F(s) = e^{-2s}[(s-1)^{-1} - (s+2)^{-1}]/3$. For ease in calculations
     let us define $G(s) = (s-1)^{-1}$ and $H(s) = (s+2)^{-1}$. Then
     $F(s) = [e^{-2s}G(s) - e^{-2s}H(s)]/3$. Using the fact that
     $\mathcal{L}\{e^{at}\} = (s-a)^{-1}$ and applying Theorem 6.3.1, we have
     $F(s) = [e^{-2s}\mathcal{L}\{e^t\} - e^{-2s}\mathcal{L}\{e^{-2t}\}]/3$. Thus
     $F(s) = [\mathcal{L}\{u_2(t)e^{(t-2)}\} - \mathcal{L}\{u_2(t)e^{-2(t-2)}\}]/3$. Using the
     linearity of the Laplace transform, we have
     $\mathcal{L}\{f(t)\} = \mathcal{L}\{u_2(t)[e^{t-2} - e^{-2(t-2)}]/3\}$. Hence,
     $f(t) = [u_2(t)(e^{t-2} - e^{-2(t-2)})]/3$. An alternate method is
     to complete the square in the denominator:
     $$F(s) = \frac{e^{-2s}}{(s+1/2)^2 - 9/4}.$$ This gives
     $f(t) = (2/3)u_2(t)e^{-(t-2)/2}\sinh\frac{3}{2}(t-2)$, which can be shown
     to be the same as that found above.

21. By completing the square in the denominator of F we can
     write $F(s) = (2s+1)/[(2s+1)^2 + 4]$. This has the form
     $G(2s+1)$ where $G(u) = u/(u^2+4)$. We must find
     $\mathcal{L}^{-1}\{G(2s+1)\}$. Applying the results of Problem 19(c), we
     have $\mathcal{L}^{-1}\{F(s)\} = \frac{1}{2}e^{-t/2}\cos(\frac{2t}{2})$.

22. If the approach of Problem 21 is used we find $f(t) = (1/3)e^{2t/3}\sinh(t/3)$, which is equivalent to the given answer using the definition of sinh t.

27. Assuming that term-by-term integration of the infinite series is permissible and recalling that $\mathcal{L}\{u_c(t)\} = e^{-cs}/s$ for $s > 0$, we have $\mathcal{L}\{f(t)\} = (1/s) + \sum_{k=1}^{\infty}(-1)^k\mathcal{L}\{u_k(t)\}$

$= (1/s) + \sum_{k=1}^{\infty}(-1)^k e^{-ks}/s = [\sum_{k=0}^{\infty}(-e^{-s})^k]/s$. We recognize

the last infinite series as the geometric series, $\sum_{k=0}^{\infty}ar^k$,

with $r = -e^{-s}$. This series converges to $[1/(1+e^{-s})]$ if $|r| < 1$. Hence, $\mathcal{L}\{f(t)\} = (1/s)[1/(1+e^{-s})]$, $s > 0$.

28. Using the definition of the Laplace transform we have $F(s) = \mathcal{L}\{f(t)\} = \int_0^{\infty}e^{-st}f(t)dt$. Since f is periodic with period T, we have $f(t+T) = f(t)$. This suggests that we rewrite the improper integral as $\int_0^{\infty}e^{-st}f(t)dt =$

$\sum_{n=0}^{\infty}\int_{nT}^{(n+1)T}e^{-st}f(t)dt$. The periodicity of f also suggests that we make the change of variable $t = r + nT$. Hence,

$F(s) = \sum_{n=0}^{\infty}\int_0^T e^{-s(r+nT)}f(r+nT)dr = \sum_{n=0}^{\infty}(e^{-sT})^n \cdot \int_0^T e^{-rs}f(r)dr$,

where we have used the fact that $f(r+nT) = f(r+(n-1)T) = \ldots = f(r+T) = f(r)$ from the definition that f is periodic. We recognize this last

series as the geometric series, $\sum_{n=0}^{\infty}au^n$, with

$a = \int_0^T e^{-rs}f(r)dr$ and $u = e^{-sT}$. The geometric series converges to $a/(1-u)$ for $|u| < 1$ and consequently we obtain

$F(s) = (1 - e^{-sT})^{-1}\int_0^T e^{-rs}f(r)dr$, $s > 0$.

30. The function f is periodic with period 2. The result of Problem 28 gives us $\mathcal{L}\{f(t)\} = \int_0^2 e^{-st}f(t)dt/(1-e^{-2s})$.

Calculating the integral we have

$$\int_0^2 e^{-st} f(t)\,dt = \int_0^1 e^{-st}\,dt - \int_1^2 e^{-st}\,dt$$

$$= (1-e^{-s})/s + (e^{-2s}-e^{-s})/s$$

$$= (e^{-2s}-2e^{-s}+1)/s$$

$$= (1-e^{-s})^2/s. \quad \text{Since the denominator of}$$

$\mathcal{L}\{f(t)\}$, $1 - e^{-2s}$, may be written as $(1-e^{-s})(1+e^{-s})$ we obtain the desired answer.

Section 6.4, Page 318

1. f(t) can be written in the form $f(t) = 1 - u_{\pi/2}(t)$ and
    thus the Laplace transforms of the D.E. is
    $(s^2+1)Y(s) - sy(0) - y'(0) = (1/s) - e^{-\pi s/2}/s$. Introducing
    the I.C. and solving for Y(s), we obtain
    $Y(s) = (s^2+1)^{-1} + [s(s^2+1)]^{-1} - e^{-\pi s/2}/s(s^2+1)$. Using
    partial fractions on the second and third terms we find
    $Y(s)=(1/s) + (s^2+1)^{-1} - s/(s^2+1) - e^{-\pi s/2}/s + e^{-\pi s/2}s/(s^2+1)$.
    The inverse transform of the first three terms can be
    obtained directly from Table 6.2.1. Using Theorem 6.3.1
    to find the inverse transform of the last two terms we
    have $\mathcal{L}^{-1}\{e^{-\pi s/2}/s\} = u_{\pi/2}(t)g(t - \pi/2)$ where

    $g(t) = \mathcal{L}^{-1}\{1/s\} = 1$ and
    $\mathcal{L}^{-1}\{e^{-\pi s/2}s/(s^2+1)\} = u_{\pi/2}(t)h(t - \pi/2)$ where

    $h(t) = \mathcal{L}^{-1}\{s/(s^2+1)\} = \cos t$. Hence,
    $y = 1 + \sin t - \cos t + u_{\pi/2}(t)[\cos(t - \pi/2) - 1]$
    $= 1 + \sin t - \cos t - u_{\pi/2}(t)[1 - \sin t]$. The graph of the
    forcing function is a unit pulse for $0 \le t < \pi/2$ and 0
    thereafter. The graph of the solution will be composed
    of two segments. The first, for $0 \le t < \pi/2$, is a
    sinusoid oscillating about 1, which represents the system
    response to a unit forcing function and the given initial
    conditions. For $t \ge \pi/2$, the forcing function, f(t), is
    zero and the "initial" conditions are
    $y(\pi/2) = \lim_{t\to\pi^-/2} 1 + \sin t - \cos t = 2$ and
    $y'(\pi/2) = \lim_{t\to\pi^-/2} \cos t + \sin t = 1$. In this case the system
    response is $y(t) = 2\sin t - \cos t$, which is a sinusoid
    oscillating about zero.

3. According to Theorem 6.3.1,
    $\mathcal{L}\{u_{2\pi}(t)\sin(t-2\pi)\} = e^{-2\pi s}\mathcal{L}\{\sin t\} = e^{-2\pi s}/(s^2+1)$.
    Transforming the D.E., we have

$(s^2+4)Y(s) - sy(0) - y'(0) = 1/(s^2+1) - e^{-2\pi s}/(s^2+1)$.
Introducing the I.C. and solving for $Y(s)$, we obtain
$Y(s) = (1-e^{-2\pi s})/(s^2+1)(s^2+4)$. We apply partial fractions
to write
$Y(s) = [s^2+1)^{-1} - (s^2+4)^{-1} - e^{-2\pi s}(s^2+1)^{-1} + e^{-2\pi s}(s^2+4)^{-1}]/3$.
We compute the inverse transform of the first two terms
directly from Table 6.2.1 after noting that
$(s^2+4)^{-1} = (1/2)[2/(s^2+4)]$. We apply Theorem 6.3.1 to the
last two terms to obtain the solution,
$y=(1/3)\{\sin t-(1/2)\sin 2t-u_{2\pi}(t)[\sin(t-2\pi)-(1/2)\sin 2(t-2\pi)]\}$.

This may be simplified using trigonometric identities to
$y = [(2\sin t - \sin 2t)(1-u_{2\pi}(t))]/6$. Note that the forcing
function is $\sin t - \sin(t-2\pi) = 0$ for $t \geq 2\pi$. The
solution is $y(t) = 2\sin t - \sin 2t$ for $0 \leq t < 2\pi$. Thus
$y(2\pi^-) = 0$ and $y'(2\pi^-) = 2\cos 2\pi - 2\cos 4\pi = 0$. Hence the
"initial" value problem for $t \geq 2\pi$ is $y'' + 4y = 0$,
$y(2\pi) = 0$, $y'(2\pi) = 0$, which has the trivial solution
$y \equiv 0$.

8. Taking the Laplace transform, applying the I.C. and using
    Theorem 6.3.1 we have $(s^2+s+5/4)Y(s) = (1-e^{-\pi s/2})/s^2$. Thus

    $$Y(s) = \frac{1-e^{-s/2}}{s^2(s^2+s+5/4)}$$

    $$= (1-e^{-\pi s/2})\left\{\frac{4/5}{s^2} - \frac{16/25}{s} + \frac{(16/25)s-4/25}{(s+1/2)^2+1}\right\}$$

    $= (1-e^{-\pi s/2})H(s)$, where we have used partial fractions and
    completed the square in the denominator of the last term.
    Since the numerator of the last term of H can be written
    as
    $\frac{16}{25}[(s+1/2) - 3/4]$, we see that

    $\mathcal{L}^{-1}\{H(s)\} = (4/25)(5t - 4 + 4e^{-t/2}\cos t - 3e^{-t/2}\sin t)$,
    which yields the desired solution. The graph of the
    forcing function is a ramp ($f(t) = t$) for $0 \leq t < \pi/2$ and
    a constant ($f(t) = \pi/2$) for $t \geq \pi/2$. The solution will
    be a damped sinusoid oscillating about the "ramp"
    $(20t-16)/25$ for $0 \leq t < \pi/2$ and oscillating about $2\pi/5$
    for $t \geq \pi/2$.

10. Note that $g(t) = \sin t - u_\pi(t)\sin t = \sin t + u_\pi(t)\sin(t-\pi)$.
    Proceeding as in Problem 8 we find
    $Y(s) = (1+e^{-\pi s})\dfrac{1}{(s^2+1)(s^2+s+5/4)}$. The correct partial

fraction expansion of the quotient is $\dfrac{as+b}{s^2+1} + \dfrac{cs+d}{s^2+s+5/4}$,

where
$a+c = 0$, $a+b+d = 0$, $(5/4)a+b+c = 0$ and $(5/4)b+d = 1$ by
equating coefficients. Solving for the constants yields
the desired solution.

16b. Taking the Laplace transform of the D.E. we obtain
$U(s^2+s/4+1) = k(e^{-3s/2}-e^{-5s/2})/s$, since the I.C. are zero.
Solving for U and using partial fractions yields

$U(s) = k(e^{-3s/2}-e^{-5s/2})(\dfrac{1}{s} - \dfrac{s+1/4}{s^2+s/4+1})$. Thus, if

$H(s) = (\dfrac{1}{s} - \dfrac{s+1/4}{s^2+s/4+1})$, then

$h(t) = 1 - e^{-t/8}(\cos\dfrac{3\sqrt{7}}{8}t + \dfrac{\sqrt{7}}{21}\sin\dfrac{3\sqrt{7}}{8}t)$ and
$u(t) = ku_{3/2}(t)h(t-3/2) - ku_{5/2}(t)h(t-5/2)$.

16c. In all cases the plot will be zero for $0 \le t < 3/2$. For
$3/2 \le t < 5/2$ the plot will be the system response
(damped sinusoid) to a step input of magnitude k. For
$t \ge 5/2$, the plot will be the system response to the
I.C. $u(5^-/2)$, $u'(5^-/2)$ with no forcing function. The
graph shown is for $k = 2$.
Varying k will just affect
the amplitude. Note that
the amplitude never reaches
2, which would be the steady
state response for the step
input $2u_{3/2}(t)$. Note also that
the solution and its derivative
are continuous at $t = 5/2$.

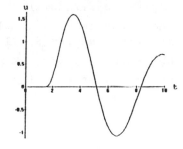

19. Since f is periodic with period 2π, we can apply the
result of Problem 28 in Section 6.3 to obtain

$\mathcal{L}\{f(t)\} = \displaystyle\int_0^{2\pi} e^{-st}f(t)\,dt / (1 - e^{-2\pi s}) = 1/s(1 + e^{-\pi s})$. Thus,

the transformed equation is
$(s^2+1)Y(s) - sy(0) - y'(0) = [s(1 + e^{-\pi s})]^{-1}$. Introducing
the I.C. and solving gives

$Y(s) = \dfrac{s}{s^2+1} + \dfrac{1}{(s^2+1)s(1+e^{-\pi s})} = \dfrac{s}{s^2+1} + \dfrac{1}{(1+e^{-\pi s})}[\dfrac{1}{s} - \dfrac{s}{s^2+1}]$

where partial fractions have been used to get the terms
in square brackets. The inverse transform of this,
however, cannot be found in Table 6.2.1. Recall that

Problem 27, Section 6.3, gave a result that had $(1 + e^{-s})$ in the denominator. This indicates that we should attempt to write the second term in the expression for $Y(s)$ as a series. In this case we can write $[1 + e^{-\pi s}]^{-1}$ as the sum of the geometric series,

$$[1 + e^{-\pi s}]^{-1} = \sum_{n=0}^{\infty} (-e^{-\pi s})^n = 1 + \sum_{n=1}^{\infty} (-e^{-\pi s})^n. \quad \text{Thus}$$

$$Y(s) = s/(s^2+1) + [1 + \sum_{n=1}^{\infty} (-e^{-\pi s})^n][(1/s) - s/(s^2+1)]$$

$$= (1/s) + \sum_{n=1}^{\infty} (-1)^n e^{-n\pi s} [(1/s) - s/(s^2+1)]. \quad \text{Assuming}$$

that term-by-term inversion of the infinite series is permissible, we apply Theorem 6.3.1 and Table 6.2.1 to obtain $y = 1 + \sum_{n=1}^{\infty} (1)^n u_{n\pi}(t) [1 - \cos(t-n\pi)]$.

Section 6.5, Page 324

1. Proceeding as in Example 1, we take the Laplace transform of the D.E. and apply the I.C.:
    $(s^2 + 2s + 2)Y(s) = s + 2 + e^{-\pi s}$. Thus,
    $Y(s) = (s+2)/[(s+1)^2 + 1] + e^{-\pi s}/[(s+1)^2 + 1]$. We write
    the first term as $(s+1)/[(s+1)^2 + 1] + 1/[(s+1)^2 + 1]$.
    Applying Theorem 6.3.1 and using Table 6.2.1, we obtain
    the solution, $y = e^{-t}\cos t + e^{-t}\sin t - u_\pi(t)e^{(t-\pi)}\sin t$. Note
    that $\sin(t-\pi) = -\sin t$.

3. Taking the Laplace transform and using the I.C. we have
    $(s^2 + 3s+2)Y(s) = \dfrac{1}{2} + e^{-5s} + \dfrac{e^{-10s}}{s}$. Thus

    $$Y(s) = \frac{1/2}{s^2+3s+2} + \frac{e^{-5s}}{s^2+3s+2} + e^{-10s}(\frac{1/2}{s}+\frac{1/2}{s+2}-\frac{1}{s+1}) \quad \text{and}$$

    hence
    $$y(t) = \frac{1}{2}h(t) + u_5(t)h(t-5) + u_{10}(t)[\frac{1}{2}+\frac{1}{2}e^{-2(t-10)}-e^{-(t-10)}]$$
    where $h(t) = e^{-t} - e^{-2t}$.

5. The Laplace transform of the D.E. is
    $(s^2+2s+3)Y(s) = \dfrac{1}{s^2+1} + e^{-3\pi s}$, so

$$Y(s) = \frac{1}{(s^2+1)(s^2+2s+3)} + e^{-3\pi s}[\frac{1}{s^2+2s+3}]. \text{ Using partial}$$

fractions or a computer algebra system we obtain

$$y(t) = \frac{1}{4}\sin t - \frac{1}{4}\cos t + \frac{1}{4}e^{-t}\cos\sqrt{2}\,t + \frac{1}{\sqrt{2}}u_{3\pi}(t)h(t-3\pi),$$

where $h(t) = e^{-t}\sin\sqrt{2}\,t$.

7. Taking the Laplace transform of the D.E. yields

$$(s^2+1)Y(s) - y'(0) = \int_0^t e^{-st}\delta(t-2\pi)\cos t\, dt. \text{ Since}$$

$\delta(t-2\pi) = 0$ for $t \neq 2\pi$ the integral on the right is equal

to $\int_{-\infty}^{\infty} e^{-st}\delta(t-2\pi)\cos t\, dt$ which equals $e^{-2\pi s}\cos 2\pi$ from

Eq.(16). Substituting for $y'(0)$ and solving for $Y(s)$

gives $Y(s) = \dfrac{1}{s^2+1} + \dfrac{e^{-2\pi s}}{s^2+1}$ and hence

$$y(t) = \sin t + u_{2\pi}(t)\sin(t-2\pi)$$

$$= \begin{cases} \sin t & 0 \leq t < 2\pi \\ 2\sin t & 2\pi \leq t \end{cases}$$

10. See the solution for Problem 7.

13a. From Eq. (22) $y(t)$ will complete one cycle when
$\sqrt{15}\,(t-5)/4 = 2\pi$ or $T = t - 5 = 8\pi/\sqrt{15}$, which is
consistent with the plot in Fig. 6.5.3. Since an impulse
causes a discontinuity in the first derivative, we need
to find the value of y' at $t = 5$ and $t = 5 + T$. From Eq.
(22) we have

$$y' = e^{-(t-5)/4}[\frac{-1}{2\sqrt{15}}\sin\frac{\sqrt{15}}{4}(t-5) + \frac{1}{2}\cos\frac{\sqrt{15}}{4}(t-5)]. \text{ Thus}$$

$y'(5) = \dfrac{1}{2}$ and $y'(5+T) = \dfrac{1}{2}e^{-T/4}$. Since the original

impulse, $\delta(t-5)$, caused a discontinuity in y' of $1/2$, we
must choose the impulse at $t = 5 + T$ to be $-e^{-T/4}$, which
is equal and opposite to y' at $5 + T$.

13b. Now consider $2y'' + y' + 2y = \delta(t-5) + k\delta(t-5-T)$ with
$y(0) = 0$, $y'(0) = 0$. Using the results of Example 1 we
have

$$y(t) = \frac{2}{\sqrt{15}}u_5(t)e^{-(t-5)/4}\sin\frac{\sqrt{15}}{4}(t-5)$$

$$+ \frac{2k}{\sqrt{15}}u_{5+T}(t)e^{-(t-5-T)/4}\sin\frac{\sqrt{15}}{4}(t-5-T)$$

$$= \frac{2}{\sqrt{15}} e^{-(t-5)/4} [u_5(t) \sin\frac{\sqrt{15}}{4}(t-5) + k u_{5+T}(t) e^{T/4} \sin\frac{\sqrt{15}}{4}(t-5-T)]$$

$$= \frac{2}{\sqrt{15}} e^{-(t-5)/4} [u_5(t) + k e^{T/4} u_{5+T}(t)] \sin\frac{\sqrt{15}}{4}(t-5). \quad \text{If}$$

$y(t) \equiv 0$ for $t \geq 5 + T$, then $1 + k e^{T/4} = 0$, or $k = -e^{-T/4}$, as found in part (a).

17b. We have $(s^2+1)Y(s) = \sum_{k=1}^{10} e^{-k\pi s}$ so that $Y(s) = \sum_{k=1}^{10} \frac{e^{-ks}}{s^2+1}$ and

hence $y(t) = \sum_{k=1}^{10} u_{k\pi}(t) \sin(t-k\pi)$

$\quad = u_\pi(t) \sin(t-\pi) + u_{2\pi}(t) \sin(t-2\pi) + \ldots + u_{10\pi} \sin(t-10\pi)$.
For $0 \leq t < \pi$, $y(t) \equiv 0$. For $\pi \leq t < 2\pi$, $y(t) = \sin(t-\pi) = -\sin t$. For $2\pi \leq t < 3\pi$, $y(t) = \sin(t-\pi) + \sin(t-2\pi) = -\sin t + \sin t \equiv 0$. Due to the periodicity of $\sin t$, the solution will exhibit this behavior in alternate intervals for $0 \leq t < 10\pi$. After $t = 2\pi$ the solution remains at zero.

21b. Substituting for $f(t)$ we have

$y = \int_0^t e^{-(t-\tau)} \delta(\tau-\pi) \sin(t-\tau) d\tau$. We know that the integration variable is always less than t (the upper limit) and thus for $t < \pi$ we have $\tau < \pi$ and thus $\delta(\tau-\pi) = 0$. Hence $y = 0$ for $t < \pi$. For $t > \pi$ utilize Eq.(16).

Section 6.6, Page 331

1c. Using the format of Eqs.(2) and (3) we have

$$f*(g*h) = \int_0^t f(t-\tau)(g*h)(\tau)d\tau$$

$$= \int_0^t f(t-\tau)[\int_0^\tau g(\tau-\eta)h(\eta)d\eta]d\tau$$

$$= \int_0^t [\int_\eta^t f(t-\tau)g(\tau-\eta)d\tau]h(\eta)(d\eta).$$

The last double integral is obtained from the previous line by interchanging the order of the η and τ integrations. Making the change of variable $\omega = \tau - \eta$ on the inside integral yields

$$f*(g*h) = \int_0^t [\int_0^{t-\eta} f(t-\eta-\omega)g(\omega)d\omega]h(\eta)d\eta$$

$$= \int_0^t (f*g)(t-\eta)h(\eta)d\eta = (f*g)*h.$$

4.   It is possible to determine f(t) explicitly by using integration by parts and then find its transform F(s). However, it is much more convenient to apply Theorem 6.6.1. Let us define $g(t) = t^2$ and $h(t) = \cos 2t$. Then, $f(t) = \int_0^t g(t-\tau)h(\tau)\,d\tau$. Using Table 6.2.1, we have $G(s) = \mathscr{L}\{g(t)\} = 2/s^3$ and $H(s) = \mathscr{L}\{h(t)\} = s/(s^2+4)$. Hence, by Theorem 6.6.1, $\mathscr{L}\{f(t)\} = F(s) = G(s)H(s) = 2/s^2(s^2+4)$.

8.   As was done in Example 1 think of F(s) as the product of s^{-4} and $(s^2+1)^{-1}$ which, according to Table 6.2.1, are the transforms of $t^3/6$ and $\sin t$, respectively. Hence, by Theorem 6.6.1, the inverse transform of F(s) is $f(t) = (1/6)\int_0^t (t-\tau)^3 \sin\tau\,d\tau$.

13. We take the Laplace transform of the D.E. and apply the I.C.: $(s^2 + 2s + 2)Y(s) = \alpha/(s^2 + \alpha^2)$. Solving for Y(s), we have $Y(s) = [\alpha/(s^2+\alpha^2)][(s+1)^2 + 1]^{-1}$, where the second factor has been written in a convenient way by completing the square. Thus Y(s) is seen to be the product of the transforms of $\sin\alpha t$ and $e^{-t}\sin t$ respectively. Hence, according to Theorem 6.6.1, $y = \int_0^t e^{-(t-\tau)}\sin(t-\tau)\sin\alpha\tau\,d\tau$.

15. Proceeding as in the above problems we obtain
$$Y(s) = \frac{s}{s^2+s+5/4} + \frac{1-e^{-s}}{s(s^2+s+5/4)}$$
$$= \frac{(s+1/?) - 1/2}{(s+1/2)^2+1} + \frac{1-e^{-s}}{s} \cdot \frac{1}{(s+1/2)^2+1},$$
where the first term is obtained by completing the square in the denominator and the second term is written as the product of two terms whose inverse transforms are known, so that Theorem 6.6.1 can be used. Note that $\mathscr{L}^{-1}\{(1-e^{-s})/s\} = 1 - u_\pi(t)$. Also note that a different form of the same solution would be obtained by writing the second term as $(1-e^{-\pi s})(\frac{a}{s} + \frac{bs + c}{(s+1/2)^2+1})$ and solving for a, b and c. In this case $\mathscr{L}^{-1}\{1-e^{-s}\} = \delta(t) - \delta(t-\pi)$ from Section 6.5.

17. Taking the Laplace transform, using the I.C. and solving, we have $Y(s) = (s+3)/(s+1)(s+2) + s/(s^2+\alpha^2)(s+1)(s+2)$. As in Problem 15, there are several correct ways the second term can be treated in order to use the convolution integral. In order to obtain the desired answer, write the second term as
$$\frac{s}{s^2+\alpha^2}(\frac{a}{s+1} + \frac{b}{s+2})$$ and solve for a and b.

20. To find $\Phi(s)$ you must recognize the integral that appears in the equation as a convolution integral.

CHAPTER 7

Section 7.1, Page 340

2. As in Example 1, let $x_1 = u$ and $x_2 = u'$, then $x_1' = x_2$ and $x_2' = u'' = 3\sin t - .5x_2 - 2x_1$.

4. In this case let $x_1 = u$, $x_2 = u'$, $x_3 = u''$, and $x_4 = u'''$.

5. Let $x_1 = u$ and $x_2 = u'$; then $x_1' = x_2$ is the first of the desired pair of equations. The second equation is obtained by substituting $u'' = x_2'$, $u' = x_2$, and $u = x_1$ in the given D.E. The I.C. become $x_1(0) = u_0$, $x_2(0) = u_0'$.

8. Follow the steps outlined in Problem 7. Solve the first D.E. for x_2 to obtain $x_2 = \frac{3}{2}x_1 - \frac{1}{2}x_1'$. Substitute this into the second D.E. to obtain $x_1'' - x_1' - 2x_1 = 0$, which has the solution $x_1 = c_1 e^{2t} + c_2 e^{-t}$. Differentiating this and substituting into the above equation for x_2 yields $x_2 = \frac{1}{2}c_1 e^{2t} + 2c_2 e^{-t}$. The I.C. then give
    $c_1 + c_2 = 3$ and $\frac{1}{2}c_1 + 2c_2 = \frac{1}{2}$, which yield
    $c_1 = \frac{11}{3}$, $c_2 = -\frac{2}{3}$. Thus $x_1 = \frac{11}{3}e^{2t} - \frac{2}{3}e^{-t}$ and
    $x_2 = \frac{11}{6}e^{2t} - \frac{4}{3}e^{-t}$. Note that for large t, the second
    term in each solution vanishes and we have $x_1 \simeq \frac{11}{3}e^{2t}$ and
    $x_2 \cong \frac{11}{6}e^{2t}$, so that $x_1 \cong 2x_2$. This says that the graph
    will be asymptotic to the line $x_1 = 2x_2$ for large t.

9. Eliminating x_2 from the two D.E.
    yields $x_2 = \frac{4}{3}x_1' - \frac{5}{3}x_1$
    and $x_1'' - 2.5x_1' + x_1 = 0$.

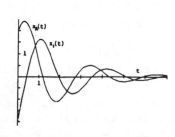

12. Solving the first D.E. for

x_2 gives $x_2 = \frac{1}{2}x_1' + \frac{1}{4}x_1$ and

substitution into the second

D.E. gives $x_1'' + x_1' + \frac{17}{4}x_1 = 0$.

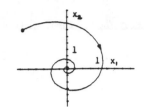

Thus $x_1 = e^{-t/2}(c_1\cos 2t + c_2\sin 2t)$ and

$x_2 = e^{-t/2}(c_2\cos 2t - c_1\sin 2t)$. The I.C. yields $c_1 = -2$ and
$c_2 = 2$.

14. If $a_{12} \neq 0$, then solve the first equation for x_2,

obtaining $x_2 = [x_1' - a_{11}x_1 - g_1(t)]/a_{12}$. Upon substituting

this expression into the second equation, we have a

second order linear O.D.E. for x_1. One I.C. is

$x_1(0) = x_1^0$. The second I.C. is

$x_2(0) = [x_1'(0) - a_{11}x_1(0) - g_1(0)]/a_{12} = x_2^0$. Solving for

$x_1'(0)$ gives $x_1'(0) = a_{12}x_2^0 + a_{11}x_1^0 + g_1(0)$. These results

hold when $a_{11}, \ldots, a_{22}$ are functions of t as long as the

derivatives exist and $a_{12}(t)$ and $a_{21}(t)$ are not both zero

on the interval. The initial conditions will involve

$a_{11}(0)$ and $a_{12}(0)$.

19. Let us number the nodes 1,2, and 3 clockwise beginning

with the top right node in Figure 7.1.4. Also let I_1,

I_2, I_3, and I_4 denote the currents through the resistor

$R = 1$, the inductor $L = 1$, the capacitor $C = \frac{1}{2}$, and the

resistor $R = 2$, respectively. Let V_1, V_2, V_3, and V_4 be

the corresponding voltage drops. Kirchhoff's first law

applied to nodes 1 and 2, respectively, gives

(i) $I_1 - I_2 = 0$ and (ii) $I_2 - I_3 - I_4 = 0$. Kirchhoff's

second law applied to each loop gives

(iii) $V_1 + V_2 + V_3 = 0$ and (iv) $V_3 - V_4 = 0$. The current-

voltage relation through each circuit element yields four

more equations: (v) $V_1 = I_1$, (vi) $I_2' = v_2$,

(vii) $(1/2)V_3' = I_3$ and (viii) $V_4 = 2I_4$. We thus have a

system of eight equations in eight unknowns, and we wish

to eliminate all of the variables except I_2 and V_3 from

this system of equations. For example, we can use

Eqs.(i) and (iv) to eliminate I_1 and V_4 in Eqs.(v) and

(viii). Then use the new Eqs.(v) and (viii) to eliminate

V_1 and I_4 in Eqs.(ii) and (iii). Finally, use the new Eqs. (ii) and (iii) in Eqs.(vi) and (vii) to obtain $I_2' = -I_2 - V_3$, $V_3' = 2I_2 - V_3$. These equations are identical (when subscripts on the remaining variables are dropped) to the equations given in the text.

21a. Note that the amount of water in each tank remains constant. Thus $Q_1(t)/30$ and $Q_2(t)/20$ represent oz./gal of salt in each tank. As in Example 3 of Section 2.5, we assume the mixture in each tank is well stirred. Then, for the first tank we have

$$\frac{dQ_1}{dt} = 1.5 - 3\frac{Q_1(t)}{30} + 1.5\frac{Q_2(t)}{20},$$ where the first term on the right represents the amount of salt per minute entering the mixture from an external source, the second term represents the loss of salt per minute going to Tank 2 and the third term represents the gain of salt per minute entering from Tank 2. Similarly, we have

$$\frac{dQ_2}{dt} = 3 + 3\frac{Q_1(t)}{30} - 4\frac{Q_2(t)}{20}$$ for Tank 2.

21b. Solve the second equation for $Q_1(t)$ to obtain $Q_1(t) = 10Q_2' + 2Q_2 - 30$. Substitution into the first equation then yields $10Q_2'' + 42Q_2' + 5Q_2 = 180$. The steady state solution for this is $Q_2^E = 180/5 = 36$. Substituting this value into the equation for Q_1 yields $Q_1^E = 72 - 30 = 42$.

21c. Substitute $Q_1 = x_1 + 42$ and $Q_2 = x_2 + 36$ into the equations found in part (a).

Section 7.2, Page 351

1c. Using Eq.(9) and following Example 1 we have
$$AB = \begin{pmatrix} 4 + 2 + 0 & -2 - 10 + 0 & 3 + 0 + 0 \\ 12 - 2 - 6 & -6 + 10 - 1 & 9 + 0 - 2 \\ -8 - 1 + 18 & 4 + 5 + 3 & -6 + 0 + 6 \end{pmatrix},$$
which yields the correct answer.

In problems 10 through 19 the method of row reduction, as illustrated in Example 2, can be used to find the inverse matrix or else to show that none exists. We start with the original matrix augmented by the indentity matrix, describe a suitable sequence of elementary row operations, and show the result of applying these operations.

10. Start with the given matrix augmented by the identity matrix.
$$\begin{pmatrix} 1 & 4 & . & 1 & 0 \\ -2 & 3 & . & 0 & 1 \end{pmatrix}$$

Add 2 times the first row to the second row.
$$\begin{pmatrix} 1 & 4 & . & 1 & 0 \\ 0 & 11 & . & 2 & 1 \end{pmatrix}$$

Multiply the second row by (1/11).
$$\begin{pmatrix} 1 & 4 & . & 1 & 0 \\ 0 & 1 & . & 2/11 & 1/11 \end{pmatrix}$$

Add (-4) times the second row to the first row.
$$\begin{pmatrix} 1 & 0 & . & 3/11 & -4/11 \\ 0 & 1 & . & 2/11 & 1/11 \end{pmatrix}$$

Since we have performed the same operation on the given matrix and the identity matrix, the 2 x 2 matric appearing on the right side of this augmented matrix is the desired inverse matrix. The answer can be checked by multiplying it by the given matrix; the result should be the indentity matrix.

12. The augmented matrix in this case is:

$$\begin{pmatrix} 1 & 2 & 3 & . & 1 & 0 & 0 \\ 2 & 4 & 5 & . & 0 & 1 & 0 \\ 3 & 5 & 6 & . & 0 & 0 & 1 \end{pmatrix}$$

Add (-2) times the first row to the second row and (-3) times the first row to the third row.

$$\begin{pmatrix} 1 & 2 & 3 & . & 1 & 0 & 0 \\ 0 & 0 & -1 & . & -2 & 1 & 0 \\ 0 & -1 & -3 & . & -3 & 0 & 1 \end{pmatrix}$$

Multiply the second and third rows by (-1) and interchange them.

$$\begin{pmatrix} 1 & 2 & 3 & . & 1 & 0 & 0 \\ 0 & 1 & 3 & . & 3 & 0 & -1 \\ 0 & 0 & 1 & . & 2 & -1 & 0 \end{pmatrix}$$

Add (-3) times the third row to the first and second rows.

$$\begin{pmatrix} 1 & 2 & 0 & . & -5 & 3 & 0 \\ 0 & 1 & 0 & . & -3 & 3 & -1 \\ 0 & 0 & 1 & . & 2 & -1 & 0 \end{pmatrix}$$

Add (-2) times the second row to the first row.

$$\begin{pmatrix} 1 & 0 & 0 & . & 1 & -3 & 2 \\ 0 & 1 & 0 & . & -3 & 3 & -1 \\ 0 & 0 & 1 & . & 2 & -1 & 0 \end{pmatrix}$$

The desired answer appears on the right side of this augmented matrix.

14. Again, start with the given matrix augmented by the identity matrix.

$$\begin{pmatrix} 1 & 2 & 1 & . & 1 & 0 & 0 \\ -2 & 1 & 8 & . & 0 & 1 & 0 \\ 1 & -2 & -7 & . & 0 & 0 & 1 \end{pmatrix}$$

Add (2) times the first row to the second row and add(-1) times the first row to the third row

$$\begin{pmatrix} 1 & 2 & 1 & . & 1 & 0 & 0 \\ 0 & 5 & 10 & . & 2 & 1 & 0 \\ 0 & -4 & -8 & . & -1 & 0 & 1 \end{pmatrix}$$

Add (4/5) times the second row to the third row.

$$\begin{pmatrix} 1 & 2 & 1 & . & 1 & 0 & 0 \\ 0 & 5 & 10 & . & 2 & 1 & 0 \\ 0 & 0 & 0 & . & 3/5 & 4/5 & 0 \end{pmatrix}$$

Since the third row of the left matrix is all zeros, no further reduction can be performed, and the given matrix is singular.

23. $\mathbf{x}' = \begin{pmatrix} 4 \\ 2 \end{pmatrix} 2e^{2t} = \begin{pmatrix} 8 \\ 4 \end{pmatrix} e^{2t}$; and

$$\begin{pmatrix} 3 & -2 \\ 2 & -2 \end{pmatrix} \mathbf{x} = \begin{pmatrix} 3 & -2 \\ 2 & -2 \end{pmatrix} \begin{pmatrix} 4 \\ 2 \end{pmatrix} e^{2t} = \begin{pmatrix} 12-4 \\ 8-4 \end{pmatrix} e^{2t} = \begin{pmatrix} 8 \\ 4 \end{pmatrix} e^{2t}.$$

26. $\Psi' = \begin{pmatrix} -3e^{-3t} & 2e^{2t} \\ 12e^{-3t} & 2e^{2t} \end{pmatrix} = \begin{pmatrix} 1 & 1 \\ 4 & -2 \end{pmatrix} \begin{pmatrix} e^{-3t} & e^{2t} \\ -4e^{-3t} & e^{2t} \end{pmatrix}.$

Section 7.3, Page 363

1. Form the augmented matrix, as in Example 1, and use row reduction.

$$\begin{pmatrix} 1 & 0 & -1 & . & 0 \\ 3 & 1 & 1 & . & 1 \\ -1 & 1 & 2 & . & 2 \end{pmatrix}$$

Add (-3) times the first row to the second and add the first row to the third.

$$\begin{pmatrix} 1 & 0 & -1 & . & 0 \\ 0 & 1 & 4 & . & 1 \\ 0 & 1 & 1 & . & 2 \end{pmatrix}$$

Add (-1) times the second row to the third.

$$\begin{pmatrix} 1 & 0 & -1 & . & 0 \\ 0 & 1 & 4 & . & 1 \\ 0 & 0 & -3 & . & 1 \end{pmatrix}$$

The third row is equivalent to $-3x_3 = 1$ or $x_3 = -1/3$. Likewise the second row is equivalent to $x_2 + 4x_3 = 1$, so $x_2 = 7/3$. Finally, from the first row, $x_1 - x_3 = 0$, so $x_1 = -1/3$. The answer can be checked by substituting into the original equations.

2. Form the augmented matrix and use row reduction to obtain

$$\begin{pmatrix} 1 & 2 & -1 & . & 1 \\ 0 & -3 & 3 & . & -1 \\ 0 & 0 & 0 & . & 1 \end{pmatrix}.$$

The last row corresponds to the equation $0x_1 + 0x_2 + 0x_3 = 1$, and there is no choice of x_1, x_2, and x_3 that satisfies this equation. Hence the given system of equations has no solution.

3. Form the augmented matrix and use row reduction.

$$\begin{pmatrix} 1 & 2 & -1 & . & 2 \\ 2 & 1 & 1 & . & 1 \\ 1 & -1 & 2 & . & -1 \end{pmatrix}$$

Add (-2) times the first row to the second and add (-1) times the first row to the third.

$$\begin{pmatrix} 1 & 2 & -1 & . & 2 \\ 0 & -3 & 3 & . & -3 \\ 0 & -3 & 3 & . & -3 \end{pmatrix}$$

Add (-1) times the second row to the third row and then multiply the second row by (-1/3).

$$\begin{pmatrix} 1 & 2 & -1 & . & 2 \\ 0 & 1 & -1 & . & 1 \\ 0 & 0 & 0 & . & 0 \end{pmatrix}$$

Since the last row has only zero entries, it may be dropped. The second row corresponds to the equation $x_2 - x_3 = 1$. We can assign an arbitrary value to either x_2 or x_3 and use this equation to solve for the other. For example, let $x_3 = c$, where c is arbitrary. Then $x_2 = 1 + c$. The first row corresponds to the equation $x_1 + 2x_2 - x_3 = 2$, so $x_1 = 2 - 2x_2 + x_3 = 2 - 2(1+c)+c = -c$.

6. To determine whether the given set of vectors is linearly independent we must solve the system $c_1\mathbf{x}^{(1)} + c_2\mathbf{x}^{(2)} + c_3\mathbf{x}^{(3)} = \mathbf{0}$ for c_1, c_2, and c_3. Form the augmented matrix and use row reduction.

$$\begin{pmatrix} 1 & 0 & 1 & . & 0 \\ 1 & 1 & 0 & . & 0 \\ 0 & 1 & 1 & . & 0 \end{pmatrix}$$

Add (-1) times the first row to the second.

$$\begin{pmatrix} 1 & 0 & 1 & . & 0 \\ 0 & 1 & -1 & . & 0 \\ 0 & 1 & 1 & . & 0 \end{pmatrix}$$

Add (-1) times the second row to the third.

$$\begin{pmatrix} 1 & 0 & 1 & . & 0 \\ 0 & 1 & -1 & . & 0 \\ 0 & 0 & 2 & . & 0 \end{pmatrix}$$

From the third row we have $c_3 = 0$. Then from the second row, $c_2 - c_3 = 0$, so $c_2 = 0$. Finally from the first row $c_1 + c_3 = 0$, so $c_1 = 0$. Since $c_1 = c_2 = c_3 = 0$, we conclude that the given vectors are linearly independent.

8.   As in Problem 6 we wish to solve the system
$c_1\mathbf{x}^{(1)} + c_2\mathbf{x}^{(2)} + c_3\mathbf{x}^{(3)} + c_4\mathbf{x}^{(4)} = \mathbf{0}$ for c_1, c_2, c_3, and c_4. Form the augmented matrix and use row reduction.

$$\begin{pmatrix} 1 & -1 & -2 & -3 & . & 0 \\ 2 & 0 & -1 & 0 & . & 0 \\ 2 & 3 & 1 & -1 & . & 0 \\ 3 & 1 & 0 & 3 & . & 0 \end{pmatrix}$$

Add (-2) times the first row to the second, add (-2) times the first row to the third, and add (-3) times the first row to the fourth.

$$\begin{pmatrix} 1 & -1 & -2 & -3 & . & 0 \\ 0 & 2 & 3 & 6 & . & 0 \\ 0 & 5 & 5 & 5 & . & 0 \\ 0 & 4 & 6 & 12 & . & 0 \end{pmatrix}$$

Multiply the second row by (1/2) and then add (-5) times the second row to the third and add (-4) times the second row to the fourth.

$$\begin{pmatrix} 1 & -1 & -2 & -3 & . & 0 \\ 0 & 1 & 3/2 & 3 & . & 0 \\ 0 & 0 & -5/2 & -10. & & 0 \\ 0 & 0 & 0 & 0 & . & 0 \end{pmatrix}$$

The third row is equivalent to the equation $c_3 + 4c_4 = 0$.
One way to satisfy this equation is by choosing $c_4 = -1$;
then $c_3 = 4$. From the second row we have
$c_2 = -(3/2)c_3 - 3c_4 = -6 + 3 = -3$. Then, from the first
row, $c_1 = c_2 + 2c_3 + 3c_4 = -3 + 8 - 3 = 2$. Hence the
given vectors are linearly dependent, and satisfy
$2\mathbf{x}^{(1)} - 3\mathbf{x}^{(2)} + 4\mathbf{x}^{(3)} - \mathbf{x}^{(4)} = \mathbf{0}$.

14. Let $t = t_0$ be a fixed value of t in the interval
$0 \le t \le 1$. To determine whether $\mathbf{x}^{(1)}(t_0)$ and $\mathbf{x}^{(2)}(t_0)$ are
linearly dependent we must solve $c_1\mathbf{x}^{(1)}(t_0) + c_2\mathbf{x}^{(2)}(t_0) = \mathbf{0}$.
We have the augmented matrix

$$\begin{pmatrix} e^{t_0} & 1 & . & 0 \\ t_0 e^{t_0} & t_0 & . & 0 \end{pmatrix}.$$

Multiply the first row by $(-t_0)$ and add to the second row
to obtain $\begin{pmatrix} e^{t_0} & 1 & . & 0 \\ 0 & 0 & . & 0 \end{pmatrix}.$

Thus, for example, we can choose $c_1 = 1$ and $c_2 = -e^{t_0}$,
and hence the given vectors are linearly dependent at t_0.
Since t_0 is arbitrary the vectors are linearly dependent
at each point in the interval. However, there is no
linear relation between $\mathbf{x}^{(1)}$ and $\mathbf{x}^{(2)}$ that is valid
throughout the interval $0 \le t \le 1$. For example, if
$t_1 \ne t_0$, and if c_1 and c_2 are chosen as above, then
$c_1\mathbf{x}^{(1)}(t_1) + c_2\mathbf{x}^{(2)}(t_1)$

$$= \begin{pmatrix} e^{t_1} \\ t_1 e^{t_1} \end{pmatrix} + -e^{t_0}\begin{pmatrix} 1 \\ t_1 \end{pmatrix} = \begin{pmatrix} e^{t_1} - e^{t_0} \\ t_1 e^{t_1} - t_1 e^{t_0} \end{pmatrix} \ne \begin{pmatrix} 0 \\ 0 \end{pmatrix}.$$

Hence the given vectors must be linearly independent on
$0 \le t \le 1$. In fact, the same argument applies to any
interval.

15. To find the eigenvalues and eigenvectors of the given

    matrix we must solve $\begin{pmatrix} 5-\lambda & -1 \\ 3 & 1-\lambda \end{pmatrix} \begin{pmatrix} x_1 \\ x_2 \end{pmatrix} = \begin{pmatrix} 0 \\ 0 \end{pmatrix}$. The

    determinant of coefficients is $(5-\lambda)(1-\lambda) - (-1)(3) = 0$,

    or $\lambda^2 - 6\lambda + 8 = 0$. Hence $\lambda_1 = 2$ and $\lambda_2 = 4$ are the

    eigenvalues. The eigenvector corresponding to λ_1 must

    satisfy $\begin{pmatrix} 3 & -1 \\ 3 & -1 \end{pmatrix} \begin{pmatrix} x_1 \\ x_2 \end{pmatrix} = \begin{pmatrix} 0 \\ 0 \end{pmatrix}$, or $3x_1 - x_2 = 0$. If we let

    $x_1 = 1$, then $x_2 = 3$ and the eigenvector is $\mathbf{x}^{(1)} = \begin{pmatrix} 1 \\ 3 \end{pmatrix}$, or

    any constant multiple of this vector. Similarly, the

    eigenvector corresponding to λ_2 must satisfy

    $\begin{pmatrix} 1 & -1 \\ 3 & -3 \end{pmatrix} \begin{pmatrix} x_1 \\ x_2 \end{pmatrix} = \begin{pmatrix} 0 \\ 0 \end{pmatrix}$, or $x_1 - x_2 = 0$. Hence $\mathbf{x}^{(2)} = \begin{pmatrix} 1 \\ 1 \end{pmatrix}$, or

    a multiple thereof.

18. The given matrix is Hermitian so we know in advance that

    its eigenvalues are real. To find the eigenvalues and

    eigenvectors we must solve $\begin{pmatrix} 1-\lambda & i \\ -i & 1-\lambda \end{pmatrix} \begin{pmatrix} x_1 \\ x_2 \end{pmatrix} = \begin{pmatrix} 0 \\ 0 \end{pmatrix}$. The

    determinant of coefficients is $(1-\lambda)^2 - i(-i) = \lambda^2 - 2\lambda$,

    so the eigenvalues are $\lambda_1 = 0$ and $\lambda_2 = 2$; observe that

    they are indeed real even though the given matrix has

    imaginary entries. The eigenvector corresponding to λ_1

    must satisfy $\begin{pmatrix} 1 & i \\ -i & 1 \end{pmatrix} \begin{pmatrix} x_1 \\ x_2 \end{pmatrix} = \begin{pmatrix} 0 \\ 0 \end{pmatrix}$, or $x_1 + ix_2 = 0$. Note

    that the second equation $-ix_1 + x_2 = 0$ is a multiple of

    the first. If $x_1 = 1$, then $x_2 = i$, and the eigenvector

    is $\mathbf{x}^{(1)} = \begin{pmatrix} 1 \\ i \end{pmatrix}$. In a similar way we find that the

    eigenvector associated with λ_2 is $\mathbf{x}^{(2)} = \begin{pmatrix} 1 \\ -i \end{pmatrix}$.

20. The eigenvalues and eigenvectors satisfy

$$\begin{pmatrix} 1-\lambda & -4 \\ 4 & -7-\lambda \end{pmatrix} \begin{pmatrix} x_1 \\ x_2 \end{pmatrix} = \begin{pmatrix} 0 \\ 0 \end{pmatrix}.$$ The determinant of coefficients

is $(1-\lambda)(-7-\lambda) - (-4)4 = \lambda^2 + 6\lambda + 9$, so $\lambda_1 = \lambda_2 = -3$.

Thus -3 is a double eigenvalue. The corresponding

eigenvectors satisfy $\begin{pmatrix} 4 & -4 \\ 4 & -4 \end{pmatrix} \begin{pmatrix} x_1 \\ x_2 \end{pmatrix} = \begin{pmatrix} 0 \\ 0 \end{pmatrix}.$ Hence

$x_1 - x_2 = 0$, so $x_1 = x_2$ and $\mathbf{x}^{(1)} = \begin{pmatrix} 1 \\ 1 \end{pmatrix}$, or any multiple

thereof. There are no other linearly independent
eigenvectors in this problem since both equations in this
last set are identical.

24. Since the given matrix is real and symmetric, we know
that the eigenvalues are real. Further, even if there
are repeated eigenvalues, there will be a full set of
three linearly independent eigenvectors. To find the
eigenvalues and eigenvectors we must solve

$$\begin{pmatrix} 3-\lambda & 2 & 4 \\ 2 & -\lambda & 2 \\ 4 & 2 & 3-\lambda \end{pmatrix} \begin{pmatrix} x_1 \\ x_2 \\ x_3 \end{pmatrix} = \begin{pmatrix} 0 \\ 0 \\ 0 \end{pmatrix}.$$ The determinant of

coefficients is $(3-\lambda)[-\lambda(3-\lambda)-4] - 2[2(3-\lambda) -8] + 4[4+4\lambda]$
$= -\lambda^3 + 6\lambda^2 + 15\lambda + 8$. Setting this equal to zero and
solving we find $\lambda_1 = \lambda_2 = -1$, $\lambda_3 = 8$. The eigenvectors
corresponding to λ_1 and λ_2 must satisfy

$$\begin{pmatrix} 4 & 2 & 4 \\ 2 & 1 & 2 \\ 4 & 2 & 4 \end{pmatrix} \begin{pmatrix} x_1 \\ x_2 \\ x_3 \end{pmatrix} = \begin{pmatrix} 0 \\ 0 \\ 0 \end{pmatrix};$$ hence there is only the single

relation $2x_1 + x_2 + 2x_3 = 0$ to be satisfied.
Consequently, two of the variables can be selected
arbitrarily and the third is then determined by this
equation. For example, if $x_1 = 1$ and $x_3 = 1$, then

$x_2 = -4$, and we obtain the eigenvector $\mathbf{x}^{(1)} = \begin{pmatrix} 1 \\ -4 \\ 1 \end{pmatrix}.$

Similarly, if $x_1 = 1$ and $x_2 = 0$, then $x_3 = -1$, and we have the eigenvector $\mathbf{x}^{(2)} = \begin{pmatrix} 1 \\ 0 \\ -1 \end{pmatrix}$, which is linearly

independent of $\mathbf{x}^{(1)}$. There are many other choices that could have been made; however, by Eq.(28) there can be no more than two linearly independent eigenvectors corresponding to the eigenvalue -1. To find the eigenvector corresponding to λ_3 we must solve

$\begin{pmatrix} -5 & 2 & 4 \\ 2 & -8 & 2 \\ 4 & 2 & -5 \end{pmatrix} \begin{pmatrix} x_1 \\ x_2 \\ x_3 \end{pmatrix} = \begin{pmatrix} 0 \\ 0 \\ 0 \end{pmatrix}$. Interchange the first and

second rows and use row reduction to obtain the equivalent system $x_1 - 4x_2 + x_3 = 0$, $2x_2 - x_3 = 0$. Since there are two equations to satisfy only one variable can be assigned an arbitrary value. If we let $x_2 = 1$, then $x_3 = 2$ and $x_1 = 2$, so we find that

$\mathbf{x}^{(3)} = \begin{pmatrix} 2 \\ 1 \\ 2 \end{pmatrix}$.

25a. From the discussion leading to Eq.(44) we find $\mathbf{T}$ is composed of the eigenvectors found in Problem 15 and thus
$\mathbf{T} = \begin{pmatrix} 1 & 1 \\ 3 & 1 \end{pmatrix}$. To verify Eq.(46) we find
$\mathbf{T}^{-1} = \begin{pmatrix} -1/2 & 1/2 \\ 3/2 & -1/2 \end{pmatrix}$ and calculate $\mathbf{T}^{-1}\mathbf{AT}$ to find that
$\mathbf{T}^{-1}\mathbf{AT} = \begin{pmatrix} 2 & 0 \\ 0 & 4 \end{pmatrix}$, as indicated in Eq.(46).

28. We are given that $\mathbf{Ax} = \mathbf{b}$ has solutions and thus we have $(\mathbf{Ax}, \mathbf{y}) = (\mathbf{b}, \mathbf{y})$. From Problem 27, though, $(\mathbf{Ax}, \mathbf{y}) = (\mathbf{x}, \mathbf{A^*y}) = 0$. Thus $(\mathbf{b}, \mathbf{y}) = 0$.

Section 7.4, Page 369

1. Use Mathematical Induction. It has already been proven that if $\mathbf{x}^{(1)}$ and $\mathbf{x}^{(2)}$ are solutions, then so is $c_1\mathbf{x}^{(1)} + c_2\mathbf{x}^{(2)}$. Assume that if $\mathbf{x}^{(1)}$, $\mathbf{x}^{(2)}$, ..., $\mathbf{x}^{(k)}$ are solutions, then $\mathbf{x} = c_1\mathbf{x}^{(1)} + \cdots + c_k\mathbf{x}^{(k)}$ is a solution.

Then use Theorem 7.4.1 to conclude that $\mathbf{x} + c_{k+1}\mathbf{x}^{(k+1)}$ is also a solution and thus $c_1\mathbf{x}^{(1)} + \cdots + c_{k+1}\mathbf{x}^{(k+1)}$ is a solution if $\mathbf{x}^{(1)}, \ldots, \mathbf{x}^{(k+1)}$ are solutions.

2a. From Eq.(10) we have
$$W = \begin{vmatrix} x_1^{(1)} & x_1^{(2)} \\ x_2^{(1)} & x_2^{(2)} \end{vmatrix} = x_1^{(1)}\, x_2^{(2)} - x_2^{(1)}\, x_1^{(2)}.$$ Taking the

derivative of these two products yields four terms which may be written as
$$\frac{dW}{dt} = [\frac{dx_1^{(1)}}{dt}x_2^{(2)} - x_2^{(1)}\frac{dx_1^{(2)}}{dt}] + [x_1^{(1)}\frac{dx_2^{(2)}}{dt} - \frac{dx_2^{(1)}}{dt}x_1^{(2)}].$$
The terms in the square brackets can now be recognized as the respective determinants appearing in the desired solution. A similar result was mentioned in Problem 20 of Section 4.1.

2b. If $\mathbf{x}^{(1)}$ is substituted into Eq.(3) we have
$$\frac{dx_1^{(1)}}{dt} = p_{11}\, x_1^{(1)} + p_{12}\, x_2^{(1)}$$
$$\frac{dx_2^{(1)}}{dt} = p_{21}\, x_1^{(1)} + p_{22}\, x_2^{(1)}.$$
Substituting the first equation above and its counterpart for $\mathbf{x}^{(2)}$ into the first determinant appearing in dW/dt

and evaluating the result yields $p_{11} \begin{vmatrix} x_1^{(1)} & x_1^{(2)} \\ x_2^{(1)} & x_2^{(2)} \end{vmatrix} = p_{11}W.$

Similarly, the second determinant in dW/dt is evaluated as $p_{22}W$, yielding the desired result.

6a. From Eq.(10) $W = \begin{vmatrix} t & t^2 \\ 1 & 2t \end{vmatrix} = 2t^2 - t^2 = t^2.$

6d. To obtain the system satisfied by $\mathbf{x}^{(1)}$ and $\mathbf{x}^{(2)}$ we consider
$$\mathbf{x} = c_1\mathbf{x}^{(1)} + c_2\mathbf{x}^{(2)}, \text{ or } \begin{pmatrix} x_1 \\ x_2 \end{pmatrix} = c_1 \begin{pmatrix} t \\ 1 \end{pmatrix} + c_2 \begin{pmatrix} t^2 \\ 2t \end{pmatrix}.$$
Taking the derivative we obtain $\begin{pmatrix} x_1' \\ x_2' \end{pmatrix} = c_1 \begin{pmatrix} 1 \\ 0 \end{pmatrix} + c_2 \begin{pmatrix} 2t \\ 2 \end{pmatrix}.$
Solving this last system for c_1 and c_2 we find $c_1 = x_1' - tx_2'$ and $c_2 = x_2'/2$. Thus

$$\begin{pmatrix} x_1 \\ x_2 \end{pmatrix} = (x_1' - tx_2') \begin{pmatrix} t \\ 1 \end{pmatrix} + \frac{x_2'}{2} \begin{pmatrix} t^2 \\ 2t \end{pmatrix}, \text{ which yields}$$

$x_1 = tx_1' - \dfrac{t^2}{2} x_2'$ and $x_2 = x_1'$. Writing this system in

matrix form we have $\mathbf{x} = \begin{pmatrix} t - t^2/2 \\ 1 \qquad 0 \end{pmatrix} \mathbf{x}'$. Finding the

inverse of the matrix multiplying $\mathbf{x}'$ yields the desired
solution.

Section 7.5, Page 378

1. Assuming that there are solutions of the form $\mathbf{x} = \xi e^{rt}$,
    we substitute into the D.E. to find

    $r\xi e^{rt} = \begin{pmatrix} 3 & -2 \\ 2 & -2 \end{pmatrix} \xi e^{rt}$. Since $\xi = I\xi = \begin{pmatrix} 1 & 0 \\ 0 & 1 \end{pmatrix} \xi$, we can

    write this equation as $\begin{pmatrix} 3 & -2 \\ 2 & -2 \end{pmatrix} \xi - r\begin{pmatrix} 1 & 0 \\ 0 & 1 \end{pmatrix} \xi = \mathbf{0}$ and

    thus we must solve $\begin{pmatrix} 3-r & -2 \\ 2 & -2-r \end{pmatrix} \begin{pmatrix} \xi_1 \\ \xi_2 \end{pmatrix} = \begin{pmatrix} 0 \\ 0 \end{pmatrix}$ for r, ξ_1, ξ_2.

    The determinant of the coefficients is
    $(3-r)(-2-r) + 4 = r^2 - r - 2$, so the eigenvalues are
    $r = -1, 2$. The eigenvector corresponding to $r = -1$
    satisfies

    $\begin{pmatrix} 4 & -2 \\ 2 & -1 \end{pmatrix} \begin{pmatrix} \xi_1 \\ \xi_2 \end{pmatrix} = \begin{pmatrix} 0 \\ 0 \end{pmatrix}$, which yields $2\xi_1 - \xi_2 = 0$. Thus

    $\mathbf{x}^{(1)}(t) = \xi^{(1)} e^{-t} = \begin{pmatrix} 1 \\ 2 \end{pmatrix} e^{-t}$, where we have set $\xi_1 = 1$.

    (Any other non zero choice would also work). In a

    similar fashion, for $r = 2$, we have $\begin{pmatrix} 1 & -2 \\ 2 & -4 \end{pmatrix} \begin{pmatrix} \xi_1 \\ \xi_2 \end{pmatrix} = \begin{pmatrix} 0 \\ 0 \end{pmatrix}$,

    or $\xi_1 - 2\xi_2 = 0$. Hence $\mathbf{x}^{(2)}(t) = \xi^{(2)} e^{2t} = \begin{pmatrix} 2 \\ 1 \end{pmatrix} e^{2t}$ by

    setting $\xi_2 = 1$. The general solution is then
    $\mathbf{x} = c_1\mathbf{x}^{(1)}(t) + c_2\mathbf{x}^{(2)}(t)$. To sketch the trajectories we
    follow the steps illustrated in Examples 1 and 2.

    Setting $c_2 = 0$ we have $\mathbf{x} = \begin{pmatrix} x_1 \\ x_2 \end{pmatrix} = c_1 \begin{pmatrix} 1 \\ 2 \end{pmatrix} e^{-t}$ or $x_1 = c_1 e^{-t}$

    and $x_2 = 2c_1 e^{-t}$ and thus one asymptote is given by

$x_2 = 2x_1$. In a similar fashion $c_1 = 0$ gives $x_2 = (1/2)x_1$ as a second asymptote. Since the roots differ in sign, the trajectories for this problem are similar in nature to those in Example 1. For $c_2 \neq 0$, all solutions will be

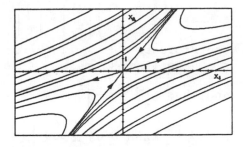

asymptotic to $x_2 = (1/2)x_1$ as $t \to \infty$. For $c_2 = 0$, the solution approaches the origin along the line $x_2 = 2x_1$.

5. Proceeding as in Problem 1 we assume a solution of the form $\mathbf{x} = \xi e^{rt}$, where r, ξ_1, ξ_2 must now satisfy

$$\begin{pmatrix} -2-r & 1 \\ 1 & -2-r \end{pmatrix} \begin{pmatrix} \xi_1 \\ \xi_2 \end{pmatrix} = \begin{pmatrix} 0 \\ 0 \end{pmatrix}.$$ Evaluating the determinant of the

coefficients set equal to zero yields $r = -1, -3$ as the eigenvalues. For $r = -1$ we find $\xi_1 = \xi_2$ and thus

$$\xi^{(1)} = \begin{pmatrix} 1 \\ 1 \end{pmatrix}$$ and for $r = -3$ we find $\xi_2 = -\xi_1$ and hence

$$\xi^{(2)} = \begin{pmatrix} 1 \\ -1 \end{pmatrix}.$$ The general solution is then

$$\mathbf{x} = c_1 \begin{pmatrix} 1 \\ 1 \end{pmatrix} e^{-t} + c_2 \begin{pmatrix} 1 \\ -1 \end{pmatrix} e^{-3t}.$$ Since there are two negative

eigenvalues, we would expect the trajectories to be similar to those of Example 2.
Setting $c_2 = 0$ and eliminating t (as in Problem 1) we find

that $\begin{pmatrix} 1 \\ 1 \end{pmatrix} e^{-t}$ approaches the

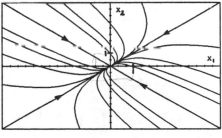

origin along the line $x_2 = x_1$.

Similarly $\begin{pmatrix} 1 \\ -1 \end{pmatrix} e^{-3t}$ approaches

the origin along the line $x_2 = -x_1$. As long as $c_1 \neq 0$, all trajectories approach the origin asymptotic to $x_2 = x_1$. For $c_1 = 0$, the trajectory approaches the origin along $x_2 = -x_1$, as shown in the graph.

7. Again assuming $\mathbf{x} = \xi e^{rt}$ we find that r, ξ_1, ξ_2 must

satisfy $\begin{pmatrix} 4-r & -3 \\ 8 & -6-r \end{pmatrix} \begin{pmatrix} \xi_1 \\ \xi_2 \end{pmatrix} = \begin{pmatrix} 0 \\ 0 \end{pmatrix}$. The determinant of the coefficients set equal to zero yields $r = 0, -2$. For $r = 0$ we find $4\xi_1 = 3\xi_2$. Choosing $\xi_2 = 4$ we find $\xi_1 = 3$ and thus $\xi^{(1)} = \begin{pmatrix} 3 \\ 4 \end{pmatrix}$. Similarly for $r = -2$ we have

$\xi^{(2)} = \begin{pmatrix} 1 \\ 2 \end{pmatrix}$ and thus $\mathbf{x} = c_1 \begin{pmatrix} 3 \\ 4 \end{pmatrix} + c_2 \begin{pmatrix} 1 \\ 2 \end{pmatrix} e^{-2t}$. To sketch the trajectories, note that the general solution is equivalent to the simultaneous equations $x_1 = 3c_1 + c_2 e^{-2t}$ and $x_2 = 4c_1 + 2c_2 e^{-2t}$. Solving the first equation for $c_2 e^{-2t}$ and substituting into the second yields $x_2 = 2x_1 - 2c_1$ and thus the trajectories are parallel straight lines.

9. The eigvalues are given by $\begin{vmatrix} 1-r & i \\ -i & 1-r \end{vmatrix} = (1-r)^2 + i^2 =$

$r(r-2) = 0$. For $r = 0$ we have $\begin{pmatrix} 1 & i \\ -i & 1 \end{pmatrix} \begin{pmatrix} \xi_1 \\ \xi_2 \end{pmatrix} = 0$ or

$-i\xi_1 + \xi_2 = 0$ and thus $\begin{pmatrix} 1 \\ i \end{pmatrix}$ is one eigenvector. Similarly

$\begin{pmatrix} 1 \\ -i \end{pmatrix}$ is the eigenvector for $r = 2$.

14. The eigenvalues and eigenvectors of the coefficient matrix satisfy $\begin{pmatrix} 1-r & -1 & 4 \\ 3 & 2-r & -1 \\ 2 & 1 & -1-r \end{pmatrix} \begin{pmatrix} \xi_1 \\ \xi_2 \\ \xi_3 \end{pmatrix} = \begin{pmatrix} 0 \\ 0 \\ 0 \end{pmatrix}$. The determinant of coefficients set equal to zero reduces to $r^3 - 2r^2 - 5r + 6 = 0$, so the eigenvalues are $r_1 = 1$, $r_2 = -2$, and $r_3 = 3$. The eigenvector corresponding to r_1 must satisfy $\begin{pmatrix} 0 & -1 & 4 \\ 3 & 1 & -1 \\ 2 & 1 & -2 \end{pmatrix} \begin{pmatrix} \xi_1 \\ \xi_2 \\ \xi_3 \end{pmatrix} = \begin{pmatrix} 0 \\ 0 \\ 0 \end{pmatrix}$.

Using row reduction we obtain the equivalent system $\xi_1 + \xi_3 = 0$, $\xi_2 - 4\xi_3 = 0$. Letting ξ_1, it follows that

$\xi_3 = -1$ and $\xi_2 = -4$, so $\xi^{(1)} = \begin{pmatrix} 1 \\ -4 \\ -1 \end{pmatrix}$. In a similar way the

eigenvectors corresponding to r_2 and r_3 are found to be

$\xi^{(2)} = \begin{pmatrix} 1 \\ -1 \\ -1 \end{pmatrix}$ and $\xi^{(3)} = \begin{pmatrix} 1 \\ 2 \\ 1 \end{pmatrix}$, respectively. Thus the

general solution of the given D.E. is

$\mathbf{x} = c_1 \begin{pmatrix} 1 \\ -4 \\ -1 \end{pmatrix} e^t + c_2 \begin{pmatrix} 1 \\ -1 \\ -1 \end{pmatrix} e^{-2t} + c_3 \begin{pmatrix} 1 \\ 2 \\ 1 \end{pmatrix} e^{3t}$. Notice that the

"trajectories" of this solution would lie in the x_1 x_2 x_3 three dimensional space.

16. The eigenvalues and eigenvectors of the coefficient

matrix are found to be $r_1 = -1$, $\xi^{(1)} = \begin{pmatrix} 1 \\ 1 \end{pmatrix}$ and $r_2 = 3$,

$\xi^{(2)} = \begin{pmatrix} 1 \\ 5 \end{pmatrix}$. Thus the general solution of the given D.E.

is $\mathbf{x} = c_1 \begin{pmatrix} 1 \\ 1 \end{pmatrix} e^{-t} + c_2 \begin{pmatrix} 1 \\ 5 \end{pmatrix} e^{3t}$. The I.C. yields the

system of equations $c_1 \begin{pmatrix} 1 \\ 1 \end{pmatrix} + c_2 \begin{pmatrix} 1 \\ 5 \end{pmatrix} = \begin{pmatrix} 1 \\ 3 \end{pmatrix}$. The augmented

matrix of this system is $\begin{pmatrix} 1 & 1 & . & 1 \\ & & . & \\ 1 & 5 & . & 3 \end{pmatrix}$ and by row reduction

we obtain $\begin{pmatrix} 1 & 1 & . & 1 \\ & & . & \\ 0 & 1 & .1/2 \end{pmatrix}$. Thus $c_2 = 1/2$ and $c_1 = 1/2$.

Substituting these values in the general solution gives the solution of the I.V.P. As $t \to \infty$, the solution

becomes asymptotic to $\mathbf{x} = \dfrac{1}{2}\begin{pmatrix} 1 \\ 5 \end{pmatrix} e^{3t}$, or $x_2 = 5x_1$.

20. Substituting $\mathbf{x} = \xi t^r$ into the D.E. we obtain

$r\xi t^r = \begin{pmatrix} 2 & -1 \\ 3 & -2 \end{pmatrix} \xi t^r$. For $t \neq 0$ this equation can be written as $\begin{pmatrix} 2-r & -1 \\ 3 & -2-r \end{pmatrix}\begin{pmatrix} \xi_1 \\ \xi_2 \end{pmatrix} = \begin{pmatrix} 0 \\ 0 \end{pmatrix}$. The eigenvalues and

eigenvectors are $r_1 = 1$, $\xi^{(1)} = \begin{pmatrix} 1 \\ 1 \end{pmatrix}$ and $r_2 = -1$,

$\xi^{(2)} = \begin{pmatrix} 1 \\ 3 \end{pmatrix}$. Substituting these in the assumed form we

obtain the general solution $\mathbf{x} = c_1\begin{pmatrix} 1 \\ 1 \end{pmatrix} t + c_2\begin{pmatrix} 1 \\ 3 \end{pmatrix} t^{-1}$.

25.
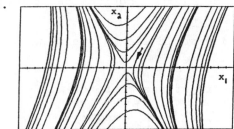
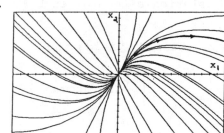

27.
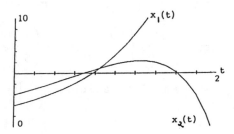

31c. The eigevalues are given by

$\begin{vmatrix} -1-r & -1 \\ -\alpha & -1-r \end{vmatrix} = r^2 + 2r + 1 - \alpha = 0.$ Thus $r_{1,2} = -1\pm\sqrt{\alpha}$.

Note that in Part (a) the eigenvalues are both negative
while in Part (b) they differ in sign. Thus, in this
part, if we choose $\alpha = 1$, then one eigenvalue is zero,
which is the transition of the one root from negative to
positive. This is the desired bifurcation point.

Section 7.6, Page 387

1.   We assume a solution of the form $\mathbf{x} = \xi e^{rt}$ thus r and ξ
are solutions of $\begin{pmatrix} 3-r & -2 \\ 4 & -1-r \end{pmatrix}\begin{pmatrix} \xi_1 \\ \xi_2 \end{pmatrix} = \begin{pmatrix} 0 \\ 0 \end{pmatrix}$. The determinant of

coefficients is $(r^2-2r-3) + 8 = r^2 - 2r + 5$, so the
eigenvalues are $r = 1 \pm 2i$. The eigenvector
corresponding to $1 + 2i$ satisfies $\begin{pmatrix} 2-2i & -2 \\ 4 & -2-2i \end{pmatrix} \begin{pmatrix} \xi_1 \\ \xi_2 \end{pmatrix} = \begin{pmatrix} 0 \\ 0 \end{pmatrix}$,

or $(2-2i)\xi_1 - 2\xi_2 = 0$. If $\xi_1 = 1$, then $\xi_2 = 1-i$ and

$\xi^{(1)} = \begin{pmatrix} 1 \\ 1-i \end{pmatrix}$ and thus one

complex-valued solution
of the D.E. is

$\mathbf{x}^{(1)}(t) = \begin{pmatrix} 1 \\ 1-i \end{pmatrix} e^{(1+2i)t}$.

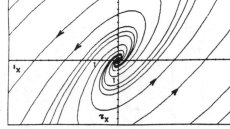

To find real-valued solutions
(see Eqs. 8 and 9) we take
the real and imaginary parts, respectively of $\mathbf{x}^{(1)}(t)$.

Thus $\mathbf{x}^{(1)}(t) = \begin{pmatrix} 1 \\ 1-i \end{pmatrix} e^t (\cos 2t + i\sin 2t)$

$$= e^t \begin{pmatrix} \cos 2t + i\sin 2t \\ \cos 2t + \sin 2t + i(\sin 2t - \cos 2t) \end{pmatrix}$$

$$= e^t \begin{pmatrix} \cos 2t \\ \cos 2t + \sin 2t \end{pmatrix} + ie^t \begin{pmatrix} \sin 2t \\ \sin 2t - \cos 2t \end{pmatrix}$$

Hence the general solution of the D.E. is

$\mathbf{x} = c_1 e^t \begin{pmatrix} \cos 2t \\ \cos 2t + \sin 2t \end{pmatrix} + c_2 e^t \begin{pmatrix} \sin 2t \\ \sin 2t - \cos 2t \end{pmatrix}$.   The

solutions spiral to ∞ as $t \to \infty$ due to the e^t terms.

7.   The eigenvalues and eigenvectors of the coefficient
matrix satisfy $\begin{pmatrix} 1-r & 0 & 0 \\ 2 & 1-r & -2 \\ 3 & 2 & 1-r \end{pmatrix} \begin{pmatrix} \xi_1 \\ \xi_2 \\ \xi_3 \end{pmatrix} = \begin{pmatrix} 0 \\ 0 \\ 0 \end{pmatrix}$.   The

determinant of coefficients reduces to $(1-r)(r^2 - 2r + 5)$
so the eigenvalues are $r_1 = 1$, $r_2 = 1 + 2i$, and
$r_3 = 1 - 2i$. The eigenvector corresponding to r_1
satisfies

$\begin{pmatrix} 0 & 0 & 0 \\ 2 & 0 & -2 \\ 3 & 2 & 0 \end{pmatrix} \begin{pmatrix} \xi_1 \\ \xi_2 \\ \xi_3 \end{pmatrix} = \begin{pmatrix} 0 \\ 0 \\ 0 \end{pmatrix}$; hence $\xi_1 - \xi_3 = 0$ and

$3\xi_1 + 2\xi_2 = 0$. If we let $\xi_2 = -3$ then $\xi_1 = 2$ and $\xi_3 = 2$, so one solution of the D.E. is $\begin{pmatrix} 2 \\ -3 \\ 2 \end{pmatrix} e^t$.   The eigenvector

corresponding to r_2 satisfies $\begin{pmatrix} -2i & 0 & 0 \\ 2 & -2i & -2 \\ 3 & 2 & -2i \end{pmatrix} \begin{pmatrix} \xi_1 \\ \xi_2 \\ \xi_3 \end{pmatrix} = \begin{pmatrix} 0 \\ 0 \\ 0 \end{pmatrix}$.

Hence $\xi_1 = 0$ and $i\xi_2 + \xi_3 = 0$.   If we let $\xi_2 = 1$, then
$\xi_3 = -i$. Thus a complex-valued solution is

$\begin{pmatrix} 0 \\ 1 \\ -i \end{pmatrix} e^t (\cos 2t + i \sin 2t)$.   Taking the real and imaginary

parts we obtain $\begin{pmatrix} 0 \\ \cos 2t \\ \sin 2t \end{pmatrix} e^t$ and $\begin{pmatrix} 0 \\ \sin 2t \\ -\cos 2t \end{pmatrix} e^t$, respectively.

Thus the general solution is

$$\mathbf{x} = c_1 \begin{pmatrix} 2 \\ -3 \\ 2 \end{pmatrix} e^t + c_2 e^t \begin{pmatrix} 0 \\ \cos 2t \\ \sin 2t \end{pmatrix} + c_3 e^t \begin{pmatrix} 0 \\ \sin 2t \\ -\cos 2t \end{pmatrix}.$$

9. The eigenvalues and eigenvectors of the coefficient

matrix satisfy $\begin{pmatrix} 2-i & -5 \\ 1 & -3-r \end{pmatrix} \begin{pmatrix} \xi_1 \\ \xi_2 \end{pmatrix} = \begin{pmatrix} 0 \\ 0 \end{pmatrix}$.   The determinant of

coefficients is $r^2 + 2r + 2$ so that the eigenvalues are
$r = -1 \pm i$.   The eigenvector corresponding to $r = -1 + i$

is given by $\begin{pmatrix} 2-i & -5 \\ 1 & -2-i \end{pmatrix} \begin{pmatrix} \xi_1 \\ \xi_2 \end{pmatrix} = \mathbf{0}$ so that $\xi_1 = (2+i)\xi_2$ and

thus one complex-valued solution is

$\mathbf{x}^{(1)}(t) = \begin{pmatrix} 2+i \\ 1 \end{pmatrix} e^{(-1+i)t}$.   Finding the real and complex

parts of $\mathbf{x}^{(1)}$ leads to the general solution

$\mathbf{x} = c_1 e^{-t} \begin{pmatrix} 2\cos t - \sin t \\ \cos t \end{pmatrix} + c_2 e^{-t} \begin{pmatrix} 2\sin t + \cos t \\ \sin t \end{pmatrix}$.   Setting

$t = 0$ we find $\mathbf{x}(0) = \begin{pmatrix} 1 \\ 1 \end{pmatrix} = c_1 \begin{pmatrix} 2 \\ 1 \end{pmatrix} + c_2 \begin{pmatrix} 1 \\ 0 \end{pmatrix}$, which is

equivalent to the system $\begin{matrix} 2c_1 + c_2 = 1 \\ c_1 + 0 = 1 \end{matrix}$.   Thus $c_1 = 1$ and

$c_2 = -1$ and

$$\mathbf{x}(t) = e^{-t}\begin{pmatrix} 2\cos t - \sin t \\ \cos t \end{pmatrix} - e^{-t}\begin{pmatrix} 2\sin t + \cos t \\ \sin t \end{pmatrix}$$

$$= e^{-t}\begin{pmatrix} \cos t - 3\sin t \\ \cos t - \sin t \end{pmatrix}, \text{ which spirals to zero as}$$

$t \to \infty$ due to the e^{-t} term.

11a. The eigenvalues are given by

$$\begin{vmatrix} 3/4-r & -2 \\ 1 & -5/4-r \end{vmatrix} = r^2 + r/2 + 17/16 = 0.$$

11d. The trajectory starts at
(5,5) in the x_1x_2 plane
and spirals around and
converges to the t axis
as $t \to \infty$.

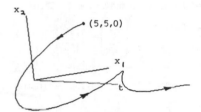

15c.

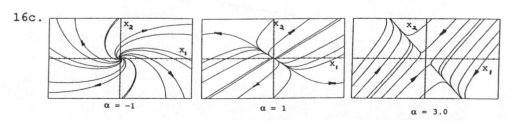

16a. $\begin{vmatrix} 5/4-r & 3/4 \\ \alpha & 5/4-r \end{vmatrix} = r^2 - 5r/2 + (25/16 - 3\alpha/4) = 0$, so

$r_{1,2} = 5/4 \pm \sqrt{3\alpha}/2$.

16b. There are two critical values of α. For $\alpha < 0$ the
eigenvalues are complex, while for $\alpha > 0$ they are real.
There will be a second critical value of α when $r_2 = 0$,
or $\alpha = 25/12$. In this case the second real eigenvalue
goes from positive to negative.

16c.

18c.

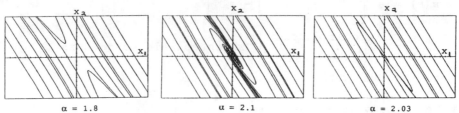

$\alpha = 1.8$ $u = 2.1$ $\alpha = 2.03$

21. If we seek solutions of the form $\mathbf{x} = \xi t^r$, then r must be an eigenvalue and ξ a corresponding eigenvector of the coefficient matrix. Thus r and ξ satisfy

$$\begin{pmatrix} -1-r & -1 \\ 2 & -1-r \end{pmatrix}\begin{pmatrix} \xi_1 \\ \xi_2 \end{pmatrix} = \begin{pmatrix} 0 \\ 0 \end{pmatrix}.$$ The determinant of coefficients

is $(-1-r)^2 +2 = r^2 + 2r + 3$, so the eigenvalues are $r = -1 \pm \sqrt{2}\,i$. The eigenvector corresponding to

$-1 \pm \sqrt{2}\,i$ satisfies $\begin{pmatrix} -\sqrt{2}\,i & -1 \\ 2 & -\sqrt{2}\,i \end{pmatrix}\begin{pmatrix} \xi_1 \\ \xi_2 \end{pmatrix} = \begin{pmatrix} 0 \\ 0 \end{pmatrix}$ or

$\sqrt{2}\,i\xi_1 + \xi_2 = 0$. If we let $\xi_1 = 1$, then $\xi_2 = -\sqrt{2}\,i$, and

$\xi^{(1)} = \begin{pmatrix} 1 \\ -\sqrt{2}\,i \end{pmatrix}$. Thus a complex-valued solution of the

given D.E. is $\begin{pmatrix} 1 \\ -\sqrt{2}\,i \end{pmatrix} t^{-1+\sqrt{2}\,i}$. From Eq. (15) of

Section 5.5 we have
$t^{-1+\sqrt{2}\,i} = t^{-1}[\cos(\sqrt{2}\,\ln t) + i\sin(\sqrt{2}\,\ln t)]$ for $t > 0$.
Separating the complex valued solution into real and imaginary parts, we obtain the two real-valued solutions

$$\mathbf{u} = t^{-1}\begin{pmatrix} \cos(\sqrt{2}\,\ln t) \\ \sqrt{2}\,\sin(\sqrt{2}\,\ln t) \end{pmatrix} \text{ and } \mathbf{v} = t^{-1}\begin{pmatrix} \sin(\sqrt{2}\,\ln t) \\ -\sqrt{2}\,\cos(\sqrt{2}\,\ln t) \end{pmatrix}.$$

23a. The eigenvalues are given by $(r+1/4)[(r+1/4)^2 + 1] = 0$.

23b.

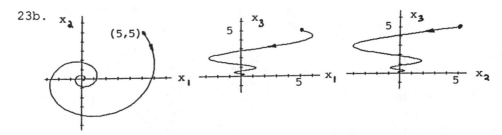

23c. Graph starts in the
first octant and
spirals around the
x_3 axis, converging
to zero.

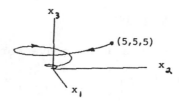

29a. We have $y_1' = x_1' = y_2$ and $y_3' = x_2' = y_4$.   Thus

$$\mathbf{y}' = \begin{pmatrix} 0 & 1 & 0 & 0 \\ -2 & 0 & 1 & 0 \\ 0 & 0 & 0 & 1 \\ 1 & 0 & -2 & 0 \end{pmatrix} \mathbf{y}.$$

29b. The eigenvalues are given by $r^4 + 4r^2 + 3 = 0$.

29c. For $r = \pm i$ the eigenvectors are given by

$$\begin{pmatrix} -i & 1 & 0 & 0 \\ -2 & -i & 1 & 0 \\ 0 & 0 & -i & 1 \\ 1 & 0 & -2 & i \end{pmatrix} \begin{pmatrix} \xi_1 \\ \xi_2 \\ \xi_3 \\ \xi_4 \end{pmatrix} = 0.$$ Choosing $\xi_1 = 1$ yields $\xi_2 = i$ and

choosing $\xi_3 = 1$ yields $\xi_4 = i$, so
$(1, i, 1, i)^T(\cos t + i\sin t)$ is a solution.   Finding the
real and imaginary parts yields
$\mathbf{w}_1 = (\cos t, -\sin t, \cos t, -\sin t)^T$ and
$\mathbf{w}_2 = (\sin t, \cos t, \sin t, \cos t)^T$ as two real solutions.
In a similar fashion, for $r = \pm \sqrt{3}\, i$, we obtain
$\mathbf{w}_3 = (\cos\sqrt{3}\,t, -\sqrt{3}\cos\sqrt{3}\,t, \sqrt{3}\sin\sqrt{3}\,t)^T$ and
$\mathbf{w}_4 = (\sin\sqrt{3}\,t, \sqrt{3}\cos\sqrt{3}\,t, -\sin\sqrt{3}\,t, -\sqrt{3}\cos\sqrt{3}\,t)^T$.
Thus $\mathbf{y} = c_1\mathbf{w}_1 + c_2\mathbf{w}_2 + c_3\mathbf{w}_3 + c_4\mathbf{w}_4$, so $\mathbf{y}^T(0) = (2, 1, 2, 1)$
yields $c_1 + c_3 = 2$, $c_2 + \sqrt{3}\,c_4 = 1$, $c_1 - c_3 = 2$, and
$c_2 - \sqrt{3}\,c_4 = 1$, which yields $c_1 = 2$, $c_2 = 1$, and

$c_3 = c_4 = 0$.   Hence $\mathbf{y} = \begin{pmatrix} 2\cos t + \sin t \\ -2\sin t + \cos t \\ 2\cos t + \sin t \\ -2\sin t + \cos t \end{pmatrix}$.

29e. The natural frequencies are $\omega_1 = 1$ and $\omega_2 = \sqrt{3}$, which
are the absolute value of the eigenvalues.   For any other
choice of I.C., both frequencies will be present, and
thus another mode of oscillation with a different
frequency (depending on the I.C.) will be present.

Section 7.7, Page 396

1. The eigenvalues and eigenvectors of the given coefficient
    matrix satisfy $\begin{pmatrix} 3-r & -4 \\ 1 & -1-r \end{pmatrix} \begin{pmatrix} \xi_1 \\ \xi_2 \end{pmatrix} = \begin{pmatrix} 0 \\ 0 \end{pmatrix}$. The determinant of
    coefficients is $(3-r)(-1-r) + 4 = r^2 - 2r + 1 = (r-1)^2$ so
    $r_1 = 1$ and $r_2 = 1$. The eigenvectors corresponding to
    this double eigenvalue satisfy $\begin{pmatrix} 2 & -4 \\ 1 & -2 \end{pmatrix} \begin{pmatrix} \xi_1 \\ \xi_2 \end{pmatrix} = \begin{pmatrix} 0 \\ 0 \end{pmatrix}$, or
    $\xi_1 - 2\xi_2 = 0$. Thus the only eigenvectors are multiples
    of $\xi^{(1)} = \begin{pmatrix} 2 \\ 1 \end{pmatrix}$. One solution of the given D.E. is
    $\mathbf{x}^{(1)}(t) = \begin{pmatrix} 2 \\ 1 \end{pmatrix} e^t$, but there is no second solution of this
    form. To find a second solution we assume, as in
    Eq. (9), that $\mathbf{x} = \eta t e^t + \zeta e^t$ and substitute this
    expression into the D.E. As in Example 1 we find that η
    is an eigenvector, so we choose $\eta = \begin{pmatrix} 2 \\ 1 \end{pmatrix}$. Then ζ must
    satisfy
    $\begin{pmatrix} 2 & -4 \\ 1 & -2 \end{pmatrix} \begin{pmatrix} \zeta_1 \\ \zeta_2 \end{pmatrix} = \begin{pmatrix} 2 \\ 1 \end{pmatrix}$, which verifies Eq.(12). Solving these
    equations yields $\zeta_1 - 2\zeta_2 = 1$. If $\zeta_2 = k$, where k is an
    arbitrary constant, then $\zeta_1 = 1 + 2k$. Hence the second
    solution that we obtain is
    $\mathbf{x}^{(2)}(t) = \begin{pmatrix} 2 \\ 1 \end{pmatrix} t e^t + \begin{pmatrix} 1 + 2k \\ k \end{pmatrix} e^t = \begin{pmatrix} 2 \\ 1 \end{pmatrix} t e^t + \begin{pmatrix} 1 \\ 0 \end{pmatrix} e^t + k \begin{pmatrix} 2 \\ 1 \end{pmatrix} e^t$.
    The last term is a multiple of the first solution $\mathbf{x}^{(1)}(t)$
    and may be neglected, that is, we may set $k = 0$. Thus
    $\mathbf{x}^{(2)}(t) = \begin{pmatrix} 2 \\ 1 \end{pmatrix} t e^t + \begin{pmatrix} 1 \\ 0 \end{pmatrix} e^t$ and the general solution is $\mathbf{x}$
    $= c_1 \mathbf{x}^{(1)}(t) + c_2 \mathbf{x}^{(2)}(t)$. All solutions diverge to
    infinity as $t \to \infty$. The graph is shown on the right.

3. The origin is attracting

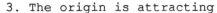

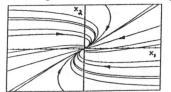

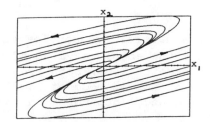

5. Substituting $\mathbf{x} = \xi e^{rt}$ into the given system, we find that
    the eigenvalues and eigenvectors satisfy

$$\begin{pmatrix} 1-r & 1 & 1 \\ 2 & 1-r & -1 \\ 0 & -1 & 1-r \end{pmatrix} \begin{pmatrix} \xi_1 \\ \xi_2 \\ \xi_3 \end{pmatrix} = \begin{pmatrix} 0 \\ 0 \\ 0 \end{pmatrix}.$$ The determinant of coefficients

is $-r^3 + 3r^2 - 4$ and thus $r_1 = -1$, $r_2 = 2$ and $r_3 = 2$.
The eigenvector corresponding to r_1 satisfies

$$\begin{pmatrix} 2 & 1 & 1 \\ 2 & 2 & -1 \\ 0 & -1 & 2 \end{pmatrix} \begin{pmatrix} \xi_1 \\ \xi_2 \\ \xi_3 \end{pmatrix} = \begin{pmatrix} 0 \\ 0 \\ 0 \end{pmatrix}$$ which yields $\xi^{(1)} = \begin{pmatrix} -3 \\ 4 \\ 2 \end{pmatrix}$ and

$\mathbf{x}^{(1)} = \begin{pmatrix} -3 \\ 4 \\ 2 \end{pmatrix} e^{-t}$. The eigenvectors corresponding to the

double eigenvalue must satsify $\begin{pmatrix} -1 & 1 & 1 \\ 2 & -1 & -1 \\ 0 & -1 & -1 \end{pmatrix} \begin{pmatrix} \xi_1 \\ \xi_2 \\ \xi_3 \end{pmatrix} = \begin{pmatrix} 0 \\ 0 \\ 0 \end{pmatrix}$,

which yields the single eigenvector $\xi^{(2)} = \begin{pmatrix} 0 \\ 1 \\ -1 \end{pmatrix}$ and hence

$\mathbf{x}^{(2)}(t) = \begin{pmatrix} 0 \\ 1 \\ -1 \end{pmatrix} e^{2t}$. The second solution corresponding to

the double eigenvalue will have the form specified by

Eq.(9), which yields $\mathbf{x}^{(3)} = \begin{pmatrix} 0 \\ 1 \\ -1 \end{pmatrix} t e^{2t} + \eta e^{2t}$.

Substituting this into the given system, or using

Eq.(12), we find that η satisfies $\begin{pmatrix} -1 & 1 & 1 \\ 2 & -1 & -1 \\ 0 & -1 & -1 \end{pmatrix} \begin{pmatrix} \eta_1 \\ \eta_2 \\ \eta_3 \end{pmatrix} = \begin{pmatrix} 0 \\ 1 \\ -1 \end{pmatrix}$.

Using row reduction we find that $\eta_1 = 1$ and $\eta_2 + \eta_3 = 1$,
where either η_2 or η_3 is arbitrary. If we choose $\eta_2 = 0$,

then $\eta = \begin{pmatrix} 1 \\ 0 \\ 1 \end{pmatrix}$ and thus $\mathbf{x}^{(3)} = \begin{pmatrix} 0 \\ 1 \\ -1 \end{pmatrix} t e^{2t} + \begin{pmatrix} 1 \\ 0 \\ 1 \end{pmatrix} e^{2t}$. The

general solution is then $\mathbf{x} = c_1 \mathbf{x}^{(1)} + c_2 \mathbf{x}^{(2)} + c_3 \mathbf{x}^{(3)}$.

9. We have $\begin{vmatrix} 2-r & 3/2 \\ -3/2 & -1-r \end{vmatrix} = (r-1/2)^2 = 0$. For $r = 1/2$, the

eigenvector is given by $\begin{pmatrix} 3/2 & 3/2 \\ -3/2 & -3/2 \end{pmatrix}\begin{pmatrix} \xi_1 \\ \xi_2 \end{pmatrix} = 0$, so $\xi = \begin{pmatrix} 1 \\ -1 \end{pmatrix}$

and $\begin{pmatrix} 1 \\ -1 \end{pmatrix}e^{t/2}$ is one solution. For the second solution we

have $\mathbf{x} = \xi t e^{t/2} + \eta e^{t/2}$, where $(\mathbf{A} - \frac{1}{2}\mathbf{I})\eta = \xi$, $\mathbf{A}$ being
the coefficient matrix for this problem. This last
equation reduces to $3\eta_1/2 - 3\eta_2/2 = 1$ and
$-3\eta_1/2 - 3\eta_2/2 = -1$. Choosing $\eta_2 = 0$ yields $\eta_1 = 2/3$ and
hence

$\mathbf{x} = c_1\begin{pmatrix} 1 \\ -1 \end{pmatrix}e^{t/2} + c_2\begin{pmatrix} 2/3 \\ 0 \end{pmatrix}e^{t/2} + c_2\begin{pmatrix} 1 \\ -1 \end{pmatrix}te^{t/2}$. $\mathbf{x}(0) = \begin{pmatrix} 3 \\ -2 \end{pmatrix}$

gives $c_1 + 2c_2/3 = 3$ and $-c_1 = -2$, and hence $c_1 = 2$,
$c_2 = 3/2$. Substituting these into the above $\mathbf{x}$ yields the
solution.

12.

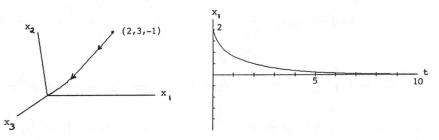

13. The eigenvalues and eigenvectors of the coefficient

matrix satisfy $\begin{pmatrix} 1-r & 1 & 1 \\ 2 & 1-r & -1 \\ -3 & 2 & 4-r \end{pmatrix}\begin{pmatrix} \xi_1 \\ \xi_2 \\ \xi_3 \end{pmatrix} = \begin{pmatrix} 0 \\ 0 \\ 0 \end{pmatrix}$. The determinant

of coefficients is $8 - 12r + 6r^2 - r^3 = (2-r)^3$, so the
eigenvalues are $r_1 = r_2 = r_3 = 2$. The eigenvectors
corresponding to this triple eigenvalue satisfy
$\begin{pmatrix} -1 & 1 & 1 \\ 2 & -1 & -1 \\ -3 & 2 & 2 \end{pmatrix}\begin{pmatrix} \xi_1 \\ \xi_2 \\ \xi_3 \end{pmatrix} = \begin{pmatrix} 0 \\ 0 \\ 0 \end{pmatrix}$. Using row reduction we can reduce

this to the equivalent system $\xi_1 - \xi_2 - \xi_3 = 0$, and
$\xi_2 + \xi_3 = 0$. If we let $\xi_2 = 1$, then $\xi_3 = -1$ and $\xi_1 = 0$,

so the only eigenvectors are multiples of $\xi = \begin{pmatrix} 0 \\ 1 \\ -1 \end{pmatrix}$. Thus

one solution of the given D.E. is $\mathbf{x}^{(1)}(t) = \begin{pmatrix} 0 \\ 1 \\ -1 \end{pmatrix} e^{2t}$, but

there are no other linearly independent solutions of this
form. We now seek a second solution of the form
$\mathbf{x} = \xi t e^{2t} + \eta e^{2t}$. As in the text, ξ must be an

eigenvector, so we choose $\xi = \begin{pmatrix} 0 \\ 1 \\ -1 \end{pmatrix}$. Then η must satisfy

a system of the form given by Eq.(20), that is

$\begin{pmatrix} -1 & 1 & 1 \\ 2 & -1 & -1 \\ -3 & 2 & 2 \end{pmatrix} \begin{pmatrix} \eta_1 \\ \eta_2 \\ \eta_3 \end{pmatrix} = \begin{pmatrix} 0 \\ 1 \\ -1 \end{pmatrix}$. By row reduction this is

equivalent to the system $\begin{pmatrix} 1 & -1 & -1 \\ 0 & 1 & 1 \\ 0 & 0 & 0 \end{pmatrix} \begin{pmatrix} \eta_1 \\ \eta_2 \\ \eta_3 \end{pmatrix} = \begin{pmatrix} 0 \\ 1 \\ 0 \end{pmatrix}$. If we

choose $\eta_2 = 0$, then $\eta_3 = 1$ and $\eta_1 = 1$, so $\eta = \begin{pmatrix} 1 \\ 0 \\ 1 \end{pmatrix}$. Hence

a second solution of the D.E. is

$\mathbf{x}^{(2)}(t) = \begin{pmatrix} 0 \\ 1 \\ -1 \end{pmatrix} t e^{2t} + \begin{pmatrix} 1 \\ 0 \\ 1 \end{pmatrix} e^{2t}$. Finally, we seek a third

solution of the form $\mathbf{x} = \xi(t^2/2)e^{2t} + \eta t e^{2t} + \zeta e^{2t}$, from
Eq.(22), where ξ and η are as above, and ζ satisfies a
system of the form given by Eq.(23), namely

$\begin{pmatrix} -1 & 1 & 1 \\ 2 & -1 & -1 \\ -3 & 2 & 2 \end{pmatrix} \begin{pmatrix} \zeta_1 \\ \zeta_2 \\ \zeta_3 \end{pmatrix} = \begin{pmatrix} 1 \\ 0 \\ 1 \end{pmatrix}$. By row reduction we find the

equivalent system $\begin{pmatrix} 1 & -1 & -1 \\ 0 & 1 & 1 \\ 0 & 0 & 0 \end{pmatrix} \begin{pmatrix} \zeta_1 \\ \zeta_2 \\ \zeta_3 \end{pmatrix} = \begin{pmatrix} -1 \\ 2 \\ 0 \end{pmatrix}$. If we let

$\zeta_2 = 0$, then $\zeta_3 = 2$ and $\zeta_1 = 1$, so $\underline{\zeta} = \begin{pmatrix} 1 \\ 0 \\ 2 \end{pmatrix}$. Since any

multiple of a solution is again a solution it is
convenient to multiply by 2 and write

$$\mathbf{x}^{(3)}(t) = \begin{pmatrix} 0 \\ 1 \\ -1 \end{pmatrix} t^2 e^{2t} + 2\begin{pmatrix} 1 \\ 0 \\ 1 \end{pmatrix} te^{2t} + 2\begin{pmatrix} 1 \\ 0 \\ 2 \end{pmatrix} e^{2t}. \text{ The general}$$

solution is then $\mathbf{x} = c_1\mathbf{x}^{(1)}(t) + c_2\mathbf{x}^{(2)}(t) + c_3\mathbf{x}^{(3)}(t)$.

15. Assuming $\mathbf{x} = \xi t^r$ and substituting into the given system,
we find r and ξ must satisfy $\begin{pmatrix} 1-r & -4 \\ 4 & -7-r \end{pmatrix}\begin{pmatrix} \xi_1 \\ \xi_2 \end{pmatrix} = \begin{pmatrix} 0 \\ 0 \end{pmatrix}$, which

has the double eigenvalue $r = -3$ and single eigenvector
$\begin{pmatrix} 1 \\ 1 \end{pmatrix}$. Hence one solution of the given D.E. is

$\mathbf{x}^{(1)}(t) = \begin{pmatrix} 1 \\ 1 \end{pmatrix} t^{-3}$. By analogy with the scalar case

considered in Section 5.5 and Example 1 of this section,
we seek a second solution of the form $\mathbf{x} = \eta t^{-3}\ln t + \zeta t^{-3}$.
Substituting this expression into the D.E. we find that η
and ζ satisfy the equations $(\mathbf{A} - 3\mathbf{I})\eta = \mathbf{0}$ and
$(\mathbf{A} - 3\mathbf{I})\zeta = \eta$, where $\mathbf{A} = \begin{pmatrix} 1 & -4 \\ 4 & -7 \end{pmatrix}$ and $\mathbf{I}$ is the identity

matrix. Thus $\eta = \begin{pmatrix} 1 \\ 1 \end{pmatrix}$, from above, and ζ is found to be

$\begin{pmatrix} 0 \\ -1/4 \end{pmatrix}$. Thus a second solution is

$\mathbf{x}^{(2)}(t) = \begin{pmatrix} 1 \\ 1 \end{pmatrix} t^{-3}\ln t + \begin{pmatrix} 1 \\ -1/4 \end{pmatrix} t^{-3}$.

17. All solutions of the given system approach zero as
$t \to \infty$ if and only if the eigenvalues of the coefficient
matrix either are real and negative or else are complex
with negative real part. Write down the determinantal
equation satisfied by the eigenvalues and determine when
the eigenvalues are as stated.

Section 7.8, Page 404

Each of the Problems 1 through 10 has been solved in one of
the previous sections. Thus a fundamental matrix for the
given systems can be readily written down. The fundamental
matrix $\Phi(t)$ satisfying $\Phi(0) = I$ can then be found, as shown
in the following problems.

4.   From Problem 4 of Section 7.5 we have the two linearly

independent solutions $\mathbf{x}^{(1)}(t) = \begin{pmatrix} 1 \\ -4 \end{pmatrix} e^{-3t}$ and

$\mathbf{x}^{(2)}(t) = \begin{pmatrix} 1 \\ 1 \end{pmatrix} e^{2t}$. Hence a fundamental matrix Ψ is given

by $\Psi(t) = \begin{pmatrix} e^{-3t} & e^{2t} \\ -4e^{-3t} & e^{2t} \end{pmatrix}$. To find the fundamental matrix

$\Phi(t)$ satisfying the I.C. $\Phi(0) = I$ we can proceed in
either of two ways. One way is to find $\Psi(0)$, invert it
to obtain $\Psi^{-1}(0)$, and then to form the product
$\Psi(t)\Psi^{-1}(0)$, which is $\Phi(t)$. Alternatively, we can find
the first column of Φ by determining the linear
combination

$c_1\mathbf{x}^{(1)}(t) + c_2\mathbf{x}^{(2)}(t)$ that satisfies the I.C. $\begin{pmatrix} 1 \\ 0 \end{pmatrix}$. This

requires that $c_1 + c_2 = 1$, $-4c_1 + c_2 = 0$, so we obtain
$c_1 = 1/5$ and $c_2 = 4/5$. Thus the first column of $\Phi(t)$ is
$\begin{pmatrix} (1/5)e^{-3t} + (4/5)e^{2t} \\ -(4/5)e^{-3t} + (4/5)e^{2t} \end{pmatrix}$. Similarly, the second column of

Φ is that linear combination of $\mathbf{x}^{(1)}(t)$ and $\mathbf{x}^{(2)}(t)$ that

satisfies the I.C. $\begin{pmatrix} 0 \\ 1 \end{pmatrix}$. Thus we must have

$c_1 + c_2 = 0$, $-4c_1 + c_2 = 1$; therefore $c_1 = -1/5$ and
$c_2 = 1/5$. Hence the second column of $\Phi(t)$ is
$\begin{pmatrix} -(1/5)e^{-3t} + (1/5)e^{2t} \\ (4/5)e^{-3t} + (1/5)e^{2t} \end{pmatrix}$.

6.   Two linearly independent real-valued solutions of the
given D.E. were found in Problem 2 of Section 7.6. Using
the result of that problem, we have
$\Psi(t) = \begin{pmatrix} -2e^{-t}\sin 2t & 2e^{-t}\cos 2t \\ e^{-t}\cos 2t & e^{-t}\sin 2t \end{pmatrix}$. To find $\Phi(t)$
we determine the linear combinations of the columns of

$\Psi(t)$ that satisfy the I.C. $\begin{pmatrix} 1 \\ 0 \end{pmatrix}$ and $\begin{pmatrix} 0 \\ 1 \end{pmatrix}$, respectively.

In the first case c_1 and c_2 satisfy $0c_1 + 2c_2 = 1$ and $c_1 + 0c_2 = 0$. Thus $c_1 = 0$ and $c_2 = 1/2$. In the second case we have $0c_1 + 2c_2 = 0$ and $c_1 + 0c_2 = 1$, so $c_1 = 1$ and $c_2 = 0$. Using these values of c_1 and c_2 to form the first and second columns of $\Phi(t)$ respectively, we obtain

$$\Phi(t) = \begin{pmatrix} e^t\cos 2t & -2e^{-t}\sin 2t \\ (1/2)e^{-t}\sin 2t & e^{-t}\cos 2t \end{pmatrix}.$$

8. Two linearly independent solutions of this D.E. were found in Problem 1 of Section 7.7. Using that result we have $\Psi(t) = \begin{pmatrix} 2e^t & e^t + 2te^t \\ e^t & te^t \end{pmatrix}$. To find $\Phi(t)$ we determine the linear combinations of the columns of $\Psi(t)$ that satisfy the I.C. $\begin{pmatrix} 1 \\ 0 \end{pmatrix}$ and $\begin{pmatrix} 0 \\ 1 \end{pmatrix}$, respectively. In the first case we have $2c_1 + c_2 = 1$ and $c_1 + 0c_2 = 0$, so $c_1 = 0$ and $c_2 = 1$. In the second case $2c_1 + c_2 = 0$ and $c_1 + 0c_2 = 1$, so $c_1 = 1$ and $c_2 = -2$. Using these values of c_1 and c_2 to form the respective columns of $\Phi(t)$ we obtain $\Phi(t) = \begin{pmatrix} e^t + 2te^t & -4te^t \\ te^t & e^t - 2te^t \end{pmatrix}$.

14a. $A^2 = AA = \begin{pmatrix} \lambda & 1 \\ 0 & \lambda \end{pmatrix}\begin{pmatrix} \lambda & 1 \\ 0 & \lambda \end{pmatrix} = \begin{pmatrix} \lambda^2 & 2\lambda \\ 0 & \lambda^2 \end{pmatrix}$

$A^3 = AA^2 = \begin{pmatrix} \lambda & 1 \\ 0 & \lambda \end{pmatrix}\begin{pmatrix} \lambda^2 & 2\lambda \\ 0 & \lambda^2 \end{pmatrix} = \begin{pmatrix} \lambda^3 & 3\lambda^2 \\ 0 & \lambda^3 \end{pmatrix}$

14b. Based upon the results of part a, assume

$A^n = \begin{pmatrix} \lambda^n & n\lambda^{n-1} \\ 0 & \lambda^n \end{pmatrix}$, then

$A^{n+1} = AA^n = \begin{pmatrix} \lambda & 1 \\ 0 & \lambda \end{pmatrix}\begin{pmatrix} \lambda^n & n\lambda^{n-1} \\ 0 & \lambda^n \end{pmatrix}$

$= \begin{pmatrix} \lambda^{n+1} & (n+1)\lambda^n \\ 0 & \lambda^{n+1} \end{pmatrix}$, which is the same as A^n with n replaced by $n+1$. Thus, by mathematical induction, A^n has the form shown above.

14c. From Eq.(29) we have

$$\exp(\mathbf{A}t) = \mathbf{I} + \sum_{n=1}^{\infty} \frac{\mathbf{A}^n t^n}{n!}$$

$$= \mathbf{I} + \sum_{n=1}^{\infty} \begin{pmatrix} \dfrac{\lambda^n t^n}{n!} & \dfrac{n\lambda^{n-1}t^n}{n!} \\ 0 & \dfrac{\lambda^n t^n}{n!} \end{pmatrix}$$

$$= \begin{pmatrix} 1 + \sum_{n=1}^{\infty} \dfrac{\lambda^n t^n}{n!} & \sum_{n=1}^{\infty} \dfrac{\lambda^{n-1}t^n}{(n-1)!} \\ 0 & 1 + \sum_{n=1}^{\infty} \dfrac{\lambda^n t^n}{n!} \end{pmatrix}$$

$$= \begin{pmatrix} e^{\lambda t} & te^{\lambda t} \\ 0 & e^{\lambda t} \end{pmatrix}, \text{ since}$$

$$\sum_{n=1}^{\infty} \frac{\lambda^{n-1}t^n}{(n-1)!} = t(1 + \sum_{n=1}^{\infty} \frac{\lambda^n t^n}{n!}) = te^{\lambda t}.$$

14d. $\mathbf{x} = \exp(\mathbf{A}t)\mathbf{x}^0 = \begin{pmatrix} e^{\lambda t} & te^{\lambda t} \\ 0 & e^{\lambda t} \end{pmatrix}\begin{pmatrix} x_1^0 \\ x_2^0 \end{pmatrix} = \begin{pmatrix} x_1^0 e^{\lambda t} + x_2^0 te^{\lambda t} \\ x_2^0 e^{\lambda t} \end{pmatrix}$

$$= \begin{pmatrix} x_1^0 \\ x_2^0 \end{pmatrix} e^{\lambda t} + \begin{pmatrix} x_2^0 \\ 0 \end{pmatrix} te^{\lambda t}.$$

Section 7.9, Page 411

1. From Section 7.5 Problem 3 we have

   $\mathbf{x}^{(c)} = c_1 \begin{pmatrix} 1 \\ 1 \end{pmatrix} e^t + c_2 \begin{pmatrix} 1 \\ 3 \end{pmatrix} e^{-t}$. Note that

   $\mathbf{g}(t) = \begin{pmatrix} 1 \\ 0 \end{pmatrix} e^t + \begin{pmatrix} 0 \\ 1 \end{pmatrix} t$ and that $r = 1$ is an eigenvalue of

   the coefficient matrix. Thus if the method of undetermined coefficients is used, the assumed form is given by Eq.(18).

2. Using methods of previous sections, we find that the eigenvalues are $r_1 = 2$ and $r_2 = -2$, with corresponding

   eigenvectors $\begin{pmatrix} \sqrt{3} \\ 1 \end{pmatrix}$ and $\begin{pmatrix} 1 \\ -\sqrt{3} \end{pmatrix}$. Thus

$$\mathbf{x}^{(c)} = c_1 \begin{pmatrix} \sqrt{3} \\ 1 \end{pmatrix} e^{2t} + c_2 \begin{pmatrix} 1 \\ -\sqrt{3} \end{pmatrix} e^{-2t}.$$ Writing the

nonhomogeneous term as $\begin{pmatrix} 1 \\ 0 \end{pmatrix} e^t + \begin{pmatrix} 0 \\ \sqrt{3} \end{pmatrix} e^{-t}$ we see that we

can assume $\mathbf{x}^{(p)} = \mathbf{a}e^t + \mathbf{b}e^{-t}$. Substituting this in the D.E., we obtain

$$\mathbf{a}e^t - \mathbf{b}e^{-t} = \mathbf{A}\mathbf{a}e^t + \mathbf{A}\mathbf{b}e^{-t} + \begin{pmatrix} 1 \\ 0 \end{pmatrix} e^t + \begin{pmatrix} 0 \\ \sqrt{3} \end{pmatrix} e^{-t},$$ where $\mathbf{A}$

is the given coefficient matrix. All the terms involving

e^t must add to zero and thus we have $\mathbf{A}\mathbf{a} - \mathbf{a} + \begin{pmatrix} 1 \\ 0 \end{pmatrix} = \begin{pmatrix} 0 \\ 0 \end{pmatrix}$.

This is equivalent to the system
$\sqrt{3}\,a_2 = -1$ and $\sqrt{3}\,a_1 - 2a_2 = 0$, or $a_1 = -2/3$ and
$a_2 = -1/\sqrt{3}$. Likewise the terms involving e^{-t} must add

to zero, which yields $\mathbf{A}\mathbf{b} + \mathbf{b} + \begin{pmatrix} 0 \\ \sqrt{3} \end{pmatrix} = \begin{pmatrix} 0 \\ 0 \end{pmatrix}$. The solution

of this system is $b_1 = -1$ and $b_2 = 2/\sqrt{3}$. Substituting
these values for $\mathbf{a}$ and $\mathbf{b}$ into $\mathbf{x}^{(p)}$ and adding $\mathbf{x}^{(p)}$ to
$\mathbf{x}^{(c)}$ yields the desired solution.

3.   The method of undetermined coefficients is not straight
     forward since the assumed form of $\mathbf{x}^{(p)} = \mathbf{a}\cos t + \mathbf{b}\sin t$
     leads to singular equations for $\mathbf{a}$ and $\mathbf{b}$. From Problem 3
     of Section 7.6 we find that a fundamental matrix is
     $$\Psi(t) = \begin{pmatrix} 5\cos t & 5\sin t \\ 2\cos t + \sin t & -\cos t + 2\sin t \end{pmatrix}.$$ The inverse
     matrix is
     $$\Psi^{-1}(t) = \begin{pmatrix} \dfrac{\cos t - 2\sin t}{5} & \sin t \\ \dfrac{2\cos t + \sin t}{5} & -\cos t \end{pmatrix},$$ which may be found as
     in Section 7.2 or by using a computer algebra system.
     Thus we may use the method of variation of parameters
     where $\mathbf{x} = \Psi(t)\mathbf{u}(t)$ and $\mathbf{u}(t)$ is given by
     $\mathbf{u}'(t) = \Psi^{-1}(t)\mathbf{g}(t)$ from Eq.(27). For this problem
     $$\mathbf{g}(t) = \begin{pmatrix} -\cos t \\ \sin t \end{pmatrix}$$ and thus

$$\mathbf{u}'(t) = \begin{pmatrix} \dfrac{\cos t - 2\sin t}{5} & \sin t \\ \dfrac{2\cos t + \sin t}{5} & -\cos t \end{pmatrix} \begin{pmatrix} -\cos t \\ \sin t \end{pmatrix}$$

$$= \frac{1}{5} \begin{pmatrix} 2 - 3\cos 2t + \sin 2t \\ -1 - \cos 2t - 3\sin 2t \end{pmatrix},$$

after multiplying and using appropriate trigonometric identities. Integration and multiplication by Ψ yields the desired solution.

4.   In this problem we use the method illustrated in Example 1. From Problem 4 of Section 7.5 we have the

transformation matrix $\mathbf{T} = \begin{pmatrix} 1 & 1 \\ -4 & 1 \end{pmatrix}$.   Inverting $\mathbf{T}$ we find

that $\mathbf{T}^{-1} = \dfrac{1}{5} \begin{pmatrix} 1 & -1 \\ 4 & 1 \end{pmatrix}$.   If we let $\mathbf{x} = \mathbf{T}\mathbf{y}$ and substitute

into the D.E., we obtain

$$\mathbf{y}' = \frac{1}{5} \begin{pmatrix} 1 & -1 \\ 4 & 1 \end{pmatrix} \begin{pmatrix} 1 & 1 \\ 4 & -2 \end{pmatrix} \begin{pmatrix} 1 & 1 \\ -4 & 1 \end{pmatrix} \mathbf{y} + \frac{1}{5} \begin{pmatrix} 1 & -1 \\ 4 & 1 \end{pmatrix} \begin{pmatrix} e^{-2t} \\ -2e^{t} \end{pmatrix}$$

$$= \begin{pmatrix} -3 & 0 \\ 0 & 2 \end{pmatrix} \mathbf{y} + \frac{1}{5} \begin{pmatrix} e^{-2t} + 2e^{t} \\ 4e^{-2t} - 2e^{t} \end{pmatrix}.$$ This corresponds to

the two scalar equations

$$y_1' + 3y_1 = (1/5)e^{-2t} + (2/5)e^{t},$$
$$y_2' - 2y_2 = (4/5)e^{-2t} - (2/5)e^{t},$$

which may be solved by the methods of Section 2.1. For the first equation the integrating factor is e^{3t} and we obtain $(e^{3t}y_1)' = (1/5)e^{t} + (2/5)e^{4t}$, so $e^{3t}y_1 = (1/5)e^{t} + (1/10)e^{4t} + c_1$. For the second equation the integrating factor is e^{-2t}, so $(e^{-2t}y_2)' = (4/5)e^{-4t} - (2/5)e^{-t}$. Hence $e^{-2t}y_2 = -(1/5)e^{-4t} + (2/5)e^{-t} + c_2$. Thus

$$\mathbf{y} = \begin{pmatrix} 1/5 \\ -1/5 \end{pmatrix} e^{-2t} + \begin{pmatrix} 1/10 \\ 2/5 \end{pmatrix} e^{t} + \begin{pmatrix} c_1 e^{-3t} \\ c_2 e^{2t} \end{pmatrix}.$$ Finally,

multiplying by $\mathbf{T}$, we obtain

$$\mathbf{x} = \mathbf{T}\mathbf{y} = \begin{pmatrix} 0 \\ -1 \end{pmatrix} e^{-2t} + \begin{pmatrix} 1/2 \\ 0 \end{pmatrix} e^{t} + c_1 \begin{pmatrix} 1 \\ -4 \end{pmatrix} e^{-3t} + c_2 \begin{pmatrix} 1 \\ 1 \end{pmatrix} e^{2t}.$$

The last two terms are the general solution of the corresponding homogeneous system, while the first two terms constitute a particular solution of the nonhomogeneous system.

12. Since the coefficient matrix is the same as that of
    Problem 3, use the same procedure as done in that
    problem, including the Ψ^{-1} found there. In the interval
    $\pi/2 < t < \pi$ sint > 0 and cost < 0; hence $|\sin t|$ = sint,
    but $|\cos t|$ = -cost.

14. To verify that the given vector is the general solution
    of the corresponding system, it is sufficient to
    substitute it into the D.E. Note also that the two terms
    in $\mathbf{x}^{(c)}$ are linearly independent. If we seek a solution
    of the form $\mathbf{x} = \Psi(t)\mathbf{u}(t)$ then we find that the equation
    corresponding to Eq.(26) is $t\Psi(t)\mathbf{u}'(t) = \mathbf{g}(t)$, where

    $$\Psi(t) = \begin{pmatrix} t & 1/t \\ t & 3/t \end{pmatrix} \text{ and } \mathbf{g}(t) = \begin{pmatrix} 1-t^2 \\ 2t \end{pmatrix}. \quad \text{Thus}$$

    $\mathbf{u}' = (1/t)\Psi^{-1}(t)\mathbf{g}(t)$. Using a computer algebra system or
    row operations on Ψ and $\mathbf{I}$, we find that

    $$\Psi^{-1} = \begin{pmatrix} 3/2t & -1/2t \\ -t/2 & t/2 \end{pmatrix} \text{ and hence } u_1' = \frac{3}{2t^2} - \frac{3}{2} - \frac{1}{t} \text{ and}$$

    $u_2' = \dfrac{-1}{2} + \dfrac{t^2}{2} + t$, which yields $u_1 = \dfrac{-3}{2t} - \dfrac{3t}{2} - \ln t + c_1$

    and $u_2 = -\dfrac{1}{2}t + \dfrac{t^3}{6} + \dfrac{t^2}{2} + c_2$. Multiplication of $\mathbf{u}$ by

    $\Psi(t)$ yields the desired solution.

CHAPTER 8

Section 8.1, Page 421

2a. The Euler formula is $y_{n+1} = y_n + h(2y_n - t_n + 1/2)$ for
$n = 0,1,2,3$ and with $t_0 = 0$ and $y_0 = 1$. Thus
$y_1 = y_0 + .1(2y_0 - t_0 + 1/2) = 1.25$,
$y_2 = 1.25 + .1[2(1.25) - (.1) + 1/2] = 1.54$,
$y_3 = 1.54 + .1[2(1.54) - (.2) + 1/2] = 1.878$, and
$y_4 = 1.878 + .1[2(1.878) - (.3) + 1/2] = 2.2736$.

2b. Use the same formula as in Problem 2a, except now $h = .05$
and $n = 0,1...7$. Notice that only results for $n = 1,3,5$
and 7 are needed to compare with part a.

2c. Again, use the same formula as above with $h = .025$ and $n = 0,1...15$. Notice that only results for $n = 3,7,11$ and
15 are needed to compare with parts a and b.

2d. $y' = 1/2 - t + 2y$ is a first order linear D.E. Rewrite
the equation in the form $y' - 2y = 1/2 - t$ and multiply
both sides by the integrating factor e^{-2t} to obtain
$(e^{-2t}y)' = (1/2 - t)e^{-2t}$. Integrating the right side by
parts and multiplying by e^{2t} we obtain $y = ce^{-2t} + t/2$.
The I.C. $y(0) = 1 \rightarrow c = 1$ and hence the solution of the
I.V.P. is $y = \phi(x) = e^{2t} + t/2$. Thus $\phi(0.1) = 1.2714$,
$\phi(0.2) = 1.59182$, $\phi(0.3) = 1.97212$, and $\phi(0.4) = 2.42554$.

5a. The Euler formua is $y_{n+1} = y_n + h\sqrt{t_n + y_n}$ for
$n = 0,1,2...$ with $t_0 = 1$ and $y_0 = 3$. Thus
$y_1 = 3 + .05\sqrt{1+3} = 3.1$,
$y_2 = 3.1 + .05\sqrt{1.05+3.1} = 3.201858$, which is the
approximation for $y(1.1)$.

5c. The backward Euler formula is $y_{n+1} = y_n + n\sqrt{t_{n+1}+y_{n+1}}$. y_2
will be calculated to illustrate the steps. If further
values are needed, then it would be best to generalize
the procedure shown here or use an equation solver. We
have $y_1 = 3 + .05\sqrt{1.05+y_1}$, which must be solved for y_1
as follows:
$(y_1-3)^2 = (.05)^2(1.05+y_1)$ [square both sides]
$y_1^2 - 6.0025y_1 = -9 + .002625$ [collecting y_1 terms]
$(y_1-3.00125)^2 = 9.007502 - 9 + .002625 = .010127$
[completing the square and solve for y_1]
$y_1 = 3.00125 + .100633 = 3.101881$.

Likewise for y_2 we have

$y_2 = 3.101881 + .05\sqrt{1.1+y_2}$

$y_2^2 - 6.203762y_2 + 9.621666 = (.05)^2(1.1) + (.05)^2y_2$

$y_2^2 - 6.206262y_2 = -9.621666 + (.05)^2(1.1)$

$(y_2-3.103131)^2 = -9.629422 - 9.618916 = .010506$

$y_2 = 3.10313 + .102499 = 3.205630$

7a. $y_1 = y_0 + h\dfrac{y_0^2 + 2t_0y_0}{3 + t_0^2} = 2 + .05\dfrac{4 + 4}{3 + 1} = 2.10000$

$y_2 = 2.10 + .05\dfrac{4.41 + (2.1)(2.1)}{3 + (1.05)^2}$

7c. $= 2 + .05\dfrac{y_1^2 + 2(1.05)y_1}{3 + (1.05)^2}$, which is a quadratic equation

in y_1. Using the quadratic formula, or an equation
solver, we obtain $y_1 = 2.10812$. Thus

$y_2 = 2.10812 + .05\dfrac{y_2^2 + 2(1.1)y_2}{3 + (1.1)^2}$ which is again quadratic

in y_2, yielding $y_2 = 2.22506$.

9a. For part a forty steps must be taken, that is,
$n = 0,1,\ldots39$ and for part b eighty steps must taken with
$n = 0,1,\ldots79$. Thus use of a programmable calculator or
a computer is desirable.

9c. We have $y_{n+1} = y_n + h(.5-t_{n+1} + 2y_{n+1})$, which is linear in

y_{n+1} and thus we have $y_{n+1} = \dfrac{y_n + .5h - ht_{n+1}}{1 - 2h}$. Again 40

steps are needed here and 80 steps in part d. In This
case a spreadsheet is very useful. The first few and
last two lines are shown for $h = .025$:

n	y_n	t_n	y_{n+1}
0	1	0	1.06513
1	1.06513	.025	1.13303
2	1.13303	.050	1.20381
$\vdots$			
38	7.49768	.950	7.87980
39	7.87980	.975	8.28137
40	8.28137	1.000	

At least eight decimal places were used in all
calculations.

13c. Use the hint along with the values of y that you
     calculate for t between 1.6 and 1.8. Note that the slope
     (y') is positive for vlaues of y < 1.155.

16. If $y' = 1 - t + 4y$ then
     $y'' = -1 + 4y' = -1 + 4(1-t+4y) = 3 - 4t + 16y$. In
     Eq.(14) we let y_n, y'_n and y''_n denote the approximate
     values of $\phi(t_n)$, $\phi'(t_n)$, and $\phi''(t_n)$, respectively.
     Keeping the first three terms in the Taylor series we
     have
     $y_{n+1} = y_n + y'_n h + y''_n h^2/2$
     $\qquad = y_n + (1 - t_n + 4y_n)h + (3 - 4t_n + 16y_n)h^2/2$ for
     n = 0,1 with $t_0 = 0$ and $y_0 = 1$.

19. Using Eq.(6) we have $y_{n+1} = y_n + h(2y_n -1) = (1+2h)y_n - h$.
     Setting $n + 1 = k$ (and hence $n = k-1$) this becomes
     $y_k = (1 + 2h)y_{k-1} - h$, for k = 1,2,... . Since $y_0 = 1$,
     we have $y_1 = 1 + 2h - h = 1 + h = (1 + 2h)/2 + 1/2$, and
     hence $y_2 = (1 + 2h)y_1 - h = (1 + 2h)^2/2 + (1 + 2h)/2 - h$
     $\qquad\quad = (1 + 2h)^2/2 + 1/2;$
     $y_3 = (1 + 2h)y_2 - h = (1 + 2h)^3/2 + (1 + 2h)/2 - h$
     $\quad = (1 + 2h)^3/2 + 1/2$. Continuing in this fashion (or
     using induction) we obtain $y_k = (1 + 2h)^k/2 + 1/2$. For
     fixed x > 0 choose h = x/k. Then substitute for h in the
     last formula to obtain $y_k = (1 + 2x/k)^k/2 + 1/2$. Letting
     $k \to \infty$ we find $y(x) = y_k \to e^{2x}/2 + 1/2$, which is the
     exact solution. (See hint for Problem 17d.)

Section 8.2, Page 427

1.   If $y = \phi(t)$ is the exact solution of the I.V.P., then
     $\phi'(t) = 2\phi(t) - 1$ and $\phi''(t) = 2\phi'(t) = 4\phi(t) - 2$. From
     Eq.(10), $e_{n+1} = [2\phi(\bar{t}_n) - 1]h^2$, $t_n < \bar{t}_n < t_n + h$. Thus a
     bound for e_{n+1} is $|e_{n+1}| \leq [1 + 2\max_{0 \leq t \leq 1}|\phi(t)|]h^2$. Since the
     exact solution is $y = \phi(t) = [1 + \exp(2t)]/2$,
     $e_{n+1} = h^2\exp(2\bar{t}_n)$. Therefore
     $|e_1| \leq (0.1)^2\exp(0.2) = 0.012$ and
     $|e_4| \leq (0.1)^2\exp(0.8) = 0.022$, since the maximum value of
     $\exp(2\bar{t}_n)$ occurs at t = .1 and t = .4 respectively. From
     Problem 1 of Section 8.1, the actual error in the first
     step is .0107.

4. The local truncation error is $e_{n+1} = \phi''(\bar{t}_n)h^2/2$. For this
    problem $\phi'(t) = 5t - 3\phi^{1/2}(t)$ and thus
    $\phi''(t) = 5 - (3/2)\phi^{-1/2}\phi' = 19/2 - (15/2)t\phi^{-1/2}$.
    Substituting this last expression into e_{n+1} yields the
    desired answer.

7d. $e_{n+1} = -(5\pi/2)\sin(5\pi\bar{t}_n)h^2$.

8a. From Eq.(3) we have $E_n = \phi(t_n) - y_n$. Using this in
    Eq.(9) we obtain

    $E_{n+1} = E_n + h\{f[t_n,\phi(t_n)] - f(t_n,y_n)\} + \phi''(\bar{t}_n)h^2/2$. Using
    the given inequality involving L we have
    $|f[t_n,\phi(t_n)] - f(t_n,y_n)| \leq L\ |\phi(t_n) - y_n| = L|E_n|$ and thus
    $|E_{n+1}| \leq |E_n| + hL|E_n| + \max_{t_0 \leq t \leq t_n} |\phi''(t)|h^2/2 = \alpha|E_n| + \beta h^2$.

8b. Since $\alpha = 1 + hL$, $\alpha - 1 = hL$. Hence $\beta h^2(\alpha^n-1)/(\alpha-1) =$
    $\beta h^2[(1+hL)^n - 1]/hL = \beta h[(1+hL)^n - 1]/L$.

8c. $(1+hL)^n \leq \exp(nhL)$ follows from the observation that
    $\exp(nhL) = [\exp(nL)]^n = (1 + hL + h^2L^2/2! + ...)^n$.
    Noting that $nh = t_n - t_0$, the rest follows from Eq.(ii).

9. The Taylor series for $\phi(t)$ about $t = t_{n+1}$ is

    $$\phi(t) = \phi(t_{n+1}) + \phi'(t_{n+1})(t-t_{n+1}) + \phi''(t_{n+1})\frac{(t-t_{n+1})^2}{2} + ...$$

    Letting $\phi'(t) = f(t,\phi(t))$, $t = t_n$ and $h = t_{n+1} - t_n$ we

    have $\phi(t_n) = \phi(t_{n+1}) - f(t_{n+1},\phi(t_{n+1}))h + \phi''(\bar{t}_n)h^2/2$, where

    $t_n < \bar{t}_n < t_{n+1}$. Thus

    $\phi(t_{n+1}) = \phi(t_n) + f(t_{n+1},\phi(t_{n+1}))h - \phi''(\bar{t}_n)h^2/2$. Comparing
    this to Eq. 15 of Section 8.1 we then have
    $e_{n+1} = -\phi''(\bar{t}_n)h^2/2$.

Section 8.3, Page 434

1a. The improved Euler formula is
    $y_{n+1} = y_n + [y'_n + f(t_n + h, y_n + hy'_n)]h/2$ where
    $y' = f(t,y) = 2y - 1$. Hence $y'_n = 2y_n-1$, and
    $f(t_n + h, y_n + hy'_n) = 2(y_n + hy'_n) - 1$. Thus we obtain
    $y_{n+1} = y_n + [y'_n + 2(y_n + hy'_n) -1]h/2$. If desired, we can
    substitute for y'_n in terms of y_n and obtain
    $y_{n+1} = y_n + h(1+h)(2y_n-1)$, $n = 0,1,2,3$ with $y_0 = 1$. In

this case the formula for y_{n+1} was made simpler by substituting for y_n'; but this may not always be true. Thus $y_1 = 1 + .1(1.1)(1) = 1.11$ and $y_2 = 1.11 + .1(1.1)(1.22) = 1.2442$.

1b.   If $y' = 2y - 1$ then $y'' = 2y' = 2(2y-1)$.   Hence, the three-term Taylor series formula is $y_{n+1} = y_n + y_n'h + y_n''h^2/2 = y_n + y_n'h + y_n'h^2$ with $y_n' = 2y_n - 1$.   If desired, we can substitute for y_n' and obtain $y_{n+1} = y_n + h(1+h)(2y_n-1)$, $n = 0,1,2$ with $y_0 = 1$.   Thus $y_1 = 1 + .1(1.1)(1) = 1.11$ and $y_2 = 1.11 + .1(1.1)(1.22) = 1.2442$.   This problem illustrates the result that the improved Euler and three-term Taylor formulas are identical when f $(=2y-1)$ is linear in t and y.

4b.   Since $y' = 5t - 3\sqrt{y}$, we have $y'' = 5 - (3/2)y'/\sqrt{y} = (19 - 15ty^{-1/2})/2$.   Hence the three-term Taylor formula is $y_{n+1} = y_n + h(5t_n - 3y_n^{1/2}) + h^2(19-15t_ny_n^{-1/2})/4$.   Thus $y_1 = 2 + .1(-3\sqrt{2}) + .01(19)/4 = 1.62324$ and
$y_2 = 1.62324 + .1(.5-3\sqrt{1.62324}) + .01[19-15(.1)/\sqrt{1.62324}]/4$
$= 1.33557$.

5a.   The improved Euler formula is
$y_{n+1} = y_n + [y_n' + f(t_n + h, y_n + hy_n')]h/2$ where
$y' = f(t,y) = \sqrt{t+y}$.   Hence $y_n' = \sqrt{t_n+y_n}$ and
$f(t_n+h, y_n + hy_n') = \sqrt{(t_n+h)+(y_n+h\sqrt{t_n+y_n})}$.   Thus we obtain

$$y_1 = 3 + \frac{\sqrt{1+3} + \sqrt{(1.1)+(3+.1\sqrt{1+3})}}{2}(.1) = 3.20368 \text{ and}$$

$$y_2 = 3.20368 + \frac{\sqrt{1.1+3.20368} + \sqrt{(1.2)+(3.20368+.1\sqrt{1.1+3.20368})}}{2}(.1)$$
$$= 3.41478.$$

For Problems 9 through 12, since 40 steps are needed for part (c), use a computer spreadsheet, a programmable calculator or a computer program utilizing the algorithm given in the text.

14.   Since $f(t,y)$ is linear in t and y, the results should be the same as those for the improved Euler method.

15a. Since $\phi(t_n + h) = \phi(t_{n+1})$ we have, using the first part of Eq.(5) and the given equation,

$e_{n+1} = \phi(t_{n+1}) - y_{n+1} = [\phi(t_n) - y_n] + [\phi'(t_n) -$

$\dfrac{y'_n + f(t_n+h, \ y_n+hy'_n)}{2}]h + \phi''(t_n)h^2/2! + \phi'''(\bar{t}_n)h^3/3!.$

Since $y_n = \phi(t_n)$ and $y'_n = \phi'(t_n) = f(t_n,y_n)$ this reduces to

$e_{n+1} = \phi''(t_n)h^2/2! - \{f[t_n+h, \ y_n + hf(t_n,y_n)] -$

$f(t_n,y_n)\}h/2! + \phi'''(\bar{t}_n)h^3/3!$, which can be written in the form of Eq.(i).

15b. First observe that $y' = f(t,y)$ and $y'' = f_t(t,y) + f_y(t,y)y'$. Hence $\phi''(t_n) = f_t(t_n,y_n) + f_y(t_n,y_n)f(t_n,y_n)$. Using the given Taylor series, with $a = t_n$, $h = h$, $b = y_n$ and $k = hf(t_n,y_n)$ we have

$f[t_n+h,y_n+hf(t_n,y_n)] = f(t_n,y_n)+f_t(t_n,y_n)h+f_y(t_n,y_n)hf(t_n,y_n)$

$+ [f_{tt}(\xi,\eta)h^2+2f_{ty}(\xi,\eta)h^2f(t_n,y_n)+f_{yy}(\xi,\eta)h^2f^2(t_n,y_n)]/2!$

where $t_n < \xi < t_n + h$ and $|\eta-y_n| < h|f(t_n,y_n)|$. Substituting this in Eq.(i) and using the earlier expression for $\phi''(t_n)$ we find that the first term on the right side of Eq.(i) reduces to

$-[f_{tt}(\xi,\eta) + 2f_{ty}(\xi,\eta)f(t_n,y_n) + f_{yy}(\xi,\eta)f^2(t_n,y_n)]h^3/4$,

which is proportional to h^3 plus, possibly, higher order terms. The reason that there may be higher order terms is because ξ and η will, in general, depend upon h.

17. Since $\phi(t) = [4t - 3 + 19\exp(4t)]/16$ we have $\phi'''(t) = 76\exp(4t)$ and thus from Problem 15c we find

$e_{n+1} = 38[\exp(4\bar{t}_n)]h^3/3$. Thus

$|e_{n+1}| \le (38h^3/3)\exp(4) = 691.577h^3$ on $0 \le t \le 1$.

$|e_1| = |\phi(t_1) - y_1| \le (0.038/3)\exp(0.4) = 0.0188964$,

which is approximately 1/10 of the error indicated in Eq.(16) of the previous section.

23. The modified Euler formula is

$y_{n+1} = y_n + hf[t_n + h/2, \ y_n + (h/2)f(t_n,y_n)]$ where $f(t,y) = t^2 + y^2$. Since $t_0 = 0$, $y_0 = 1$ and $h = .1$, we have $y_1 = 1 + .1[(.05)^2 + (1+.05)^2] = 1.1105$ and

$y_2 = 1.1105 + .1\{(.15)^2 + [1.1105+.05(.1^2+1.1105^2)]^2\}$

$= 1.25026$.

26. Since $\phi(t_n+h) = \phi(t_{n+1})$, we have, utilizing Eq.(10) and Eq.(13), $e_{n+1} = \phi(t_{n+1}) - y_{n+1} = \phi(t_n) + \phi'(t_n)h +$

$\phi''(t_n)h^2/2! + \phi'''(\bar{t}_n)h^3/3! - y_n - hf_n - [f_t(t_n,y_n) -$

$f_y(t_n,y_n)f_n]h^2/2 = \phi'''(\bar{t}_n)h^3/3!$ where $t_n \le \bar{t}_n \le t_n + h$.

Note that we have made use of the fact that $y_n = \phi(t_n)$, which means that
$f(t_n,y_n) = f[t_n,\phi(t_n)] = \phi'(t_n)$ and
$f_t(t_n,y_n) + f_y(t_n,y_n)f_n = f_t[t_n,\phi(t_n)] + f_y[t_n,\phi(t_n)]\phi'(t_n) = \phi''(t_n)$.

27. The four-term Taylor series formula for $y' = f(t,y)$ is
$y_{n+1} = y_n + y_n'h + y_n''h^2/2! + y_n'''h^3/3!$, where $y_n' = f(t_n,y_n)$,
$y_n'' = f_t(t_n,y_n) + f_y(t_n,y_n)y_n'$, and $y_n''' = f_{tt}(t_n,y_n) +$
$f_{ty}(t_n,y_n)y_n' + [f_{yt}(t_n,y_n) + f_{yy}(t_n,y_n)y_n']y_n' +$
$f_y(t_n,y_n)y_n'' = f_{tt}(t_n,y_n) + 2f_{ty}(t_n,y_n)y_n' + f_{yy}(t_n,y_n)(y_n')^2 +$
$f_y(t_n,y_n)y_n''$. Using the theory of Taylor series with a remainder, we find the local formula error will be
$\phi''''(\bar{t}_n)h^4/4!$ where $t_n < \bar{t}_n < t_{n+1}$ and ϕ is the exact solution of the I.V.P.

Section 8.4, Page 439

2a. The Runge-Kutta formula is
$y_{n+1} = y_n + h(k_{n1} + 2k_{n2} + 2k_{n3} + k_{n4})/6$ where k_{n1}, k_{n2}
etc. are given by Eqs.(3). Thus for
$f(t,y) = 0.5 - t + 2y$, $(t_0,y_0) = (0,1)$ and $h = .1$ we have
$k_{01} = f(0,1) = .5 + 2 = 2.5$
$k_{02} = f(.05, 1.125) = .5 - .05 + 2.25 = 2.7$
$k_{03} = f(.05, 1.135) = .5 - .05 + 2.27 = 2.72$
$k_{04} = f(.1, 1.272) = .5 - .1 + 2.544 = 2.944$
and hence
$y(.1) \cong y_1 = 1 + .1(2.5 + 5.4 + 5.44 + 2.944)/6 = 1.2714$.
To approximate $y(.2)$ we have
$k_{11} = f(.1, 1.2714) = .5 - .1 + 2.5428 = 2.9428$
$k_{12} = f(.15, 1.41854) = .5 - .15 + 2.83708 = 3.18708$
$k_{13} = f(.15, 1.430754) = .5 - .15 + 2.861508 = 3.211508$
$k_{14} = f(.2, 1.5925508) = .5 - .2 + 3.1851016 = 3.4851016$
and thus
$y(.2) \cong y_2 = 1.2714 + .1(k_{11} + 2k_{12} + 2k_{13} + k_{14})/6 = 1.59182$.

5a. We have, for $h = .1$, $t_0 = 1$ and $y_0 = 3$
$k_{01} = f(t_0,y_0) = \sqrt{t_0 + y_0} = 2$
$k_{02} = f(t_0+h/2, y_0 + hk_{01}/2) = \sqrt{t_0+.05+y_0+.1} = 2.03715$
$k_{03} = f(t_0+h/2, y_0+hk_{02}/2) = \sqrt{t_0+.05+y_0+.101858} = 2.03761$
$k_{04} = f(t_0+h, y_0+hk_{03}) = \sqrt{t_0+.1+y_0-2.03761} = 2.07455$
and thus
$y_1 = 3 + .1(2 + 4.07431 + 4.07522 + 2.07455)/6 = 3.20373$.

For Problems 9 through 12 use the algorithm given in the text (or a comparable computer program), a spreadsheet, or a programmable calculator.

14a. From Problem 15 of Section 8.3 we find (note that h = αh and k = βhf)
$f[t_n + \alpha h, y_n + \beta hf(t_n,y_n)] = f(t_n,y_n) + \alpha hf_t(t_n,y_n) + \beta hf(t_n,y_n)f_y(t_n,y_n)$ + terms involving h^2. The difference between (i) and (ii) is $ht_n + [f_t(t_n,y_n)+fy(t_n,y_n)f_n]h^2/2 - h\{af(t_n,y_n) + bf[t_n + \alpha h, y_n + \beta hf(t_n,y_n)]\}$. Since $f_n = f(t_n,y_n)$, we see that if a + b = 1, bα = 1/2 and bβ = 1/2 then the difference reduces to h times the terms involving h^2.

14b. There are only three equations involving four unknowns. Choose b to be arbitrary (b = λ) and then solve for a,b, and α in terms of λ.

Section 8.5, Page 445

2a. The predictor formula is
$y_{n+1} = y_n + h(55f_n - 59f_{n-1} + 37f_{n-2} - 9f_{n-3})/24$
and the corrector formula is
$y_{n+1} = y_n + h(9f_{n+1} + 19f_n - 5f_{n-1} + f_{n-2})/24$, where
$f_n = .5 - t_n + 2y_n$. From Section 8.4, Problem 2a we have
$y_1 = 1.2714$, $y_2 = 1.59182$ and $y_3 = 1.97211$, so $f_0 = 2.5$,
$f_1 = 2.9428$, $f_2 = 3.48364$ and $f_3 = 4.14422$. Thus the predicted value of y_4 is
$y_4 = 1.97211 + (.1/24)[55(4.14422) - 59(3.48364) + 37(2.9428)-9(2.5)] = 2.4253639$. This gives
$f_4 = 4.9507278$ and thus the corrected value of $y_4 (\equiv y(.4))$ is
$y_4 = 1.97211 + (.1/24)[9(4.950728) + 19(4.14422) - 5(3.48364) + 2.9428] = 2.425532$. These results, and the next step, are summarized in the following table:

n	y_n	f_n	y_{n+1}	f_{n+1}	Corrected y_{n+1}
0	1	2.5			
1	1.2714	2.9428			
2	1.59182	3.48364			
3	1.97211	4.14422	2.425364	4.950728	2.425532
4	2.425532	4.951064	2.968070	5.936140	2.968274
5	2.968274				

where f_n is given above, y_{n+1} is given by the predictor formula, and the corrected y_{n+1} is given by the corrector formula. Note that the value for f_4 on the line for n = 4 uses the corrected value for y_4, and differs slightly from the f_4 on the line for n = 3, which uses the predicted value for y_4.

2c. The fourth order Adams-Moulton method is given by Eq. (10):

$y_{n+1} = y_n + (h/24)(9f_{n+1} + 19f_n + 5f_{n-1} + f_{n-2})$.

Substituting $f_{n+1} = .5 - t_{n+1} + 2y_{n+1}$ and solving for y_{n+1} yields

$$y_{n+1} = (\frac{4}{4-3h})[y_n + \frac{h}{24}(4.5 - 9t_{n+1} + 19f_n - 5f_{n-1} + f_{n-2})].$$

Using this formula for y_{n+1}, we then obtain:

n	y_n	f_n	y_{n+1}
0	1	2.5	
1	1.2714	2.9428	
2	1.59182	3.48364	
3	1.97211	4.14422	2.425546
4	2.425546	4.951092	2.968308
5	2.968308		

2e. We use Eq. (16):

$y_{n+1} = (1/25)(48y_n - 36y_{n-1} + 16y_{n-3} + 12hf_{n+1})$.

Substituting f_{n+1}, as in part c, and solving for y_{n+1}, we obtain

$$y_{n+1} = \frac{1}{25 - 24h}[48y_n - 36y_{n-1} + 16y_{n-3} + 12h(.5-t_{n+1})].$$

Using this formula for y_{n+1}, we then obtain:

n	y_n	f_n	y_{n+1}
0	1	2.5	
1	1.2714	2.9428	
2	1.59182	3.48364	
3	1.97211	4.14422	2.425582
4	2.425582	4.951165	2.968447
5	2.968447		

The results for y(.4) and y(.5) given in all three parts of this problem differ slightly (in the sixth decimal place) from those given in the text, since only 4 or 5 decimal places were used for the starting points y_1, y_2 and y_3.

9a. Extending the tables from Problem 2 we have for part a:

					Corrected
n	y_n	f_n	y_{n+1}	f_{n+1}	y_{n+1}
0	1	2.5			
1	1.2714	2.9428			
2	1.591818	3.483636			
3	1.972106	4.144213	2.425360	4.950720	2.425528
4	2.425528	4.951056	2.968065	5.936129	2.968269
5	2.968269	5.936537	3.619854	7.139708	3.620104
6	3.620104	7.140208	4.404883	8.609766	4.405188
7	4.405188	8.610377	5.352650	10.405300	5.353023
8	5.353023	10.406046	6.499187	12.598373	6.499642
9	6.499642	12.599284	7.888500	15.277001	7.889057
10	7.889057				

for part c:

n	y_n	f_n	y_{n+1}
0	1	2.5	
1	1.2714	2.9428	
2	1.591818	3.483636	
3	1.972106	4.144213	2.425542
4	2.425542	4.951083	2.968302
5	2.968302	5.936605	3.620166
6	3.620166	7.140332	4.405289
7	4.405289	8.610578	5.353177
8	5.353177	10.406354	6.499868
9	6.499868	12.599736	7.889379
10	7.889379		

and for part e:

n	y_n	f_n	y_{n+1}
0	1	2.5	
1	1.2714	2.9428	
2	1.591818	3.483636	
3	1.972106	4.144213	2.425578
4	2.425578	4.951156	2.968442
5	2.968442	5.936885	3.620473
6	3.620473	7.140945	4.405827
7	4.405827	8.611653	5.354027
8	5.354027	10.408054	6.501141
9	6.501141	12.602281	7.891220
10	7.891220		

Note: The Runge-Kutta method was used to generate more
accurate values for y_2 and y_3 than were used in Problem
2e. Comparing the values for $y(.4)$ and $y(.5)$ here with
those of Problem 2 will indicate the greater accuracy
here. The exact solution is $y(t) = e^{2t} + t/2$, so
$y(1) = 7.889056$.

16. Let $P_2(t) = At^2 + Bt + C$. As in Eqs. (12) and (13) let
     $P_2(t_{n-1}) = y_{n-1}$, $P_2(t_n) = y_n$, $P_2(t_{n+1}) = y_{n+1}$ and
     $P_2'(t_{n+1}) = f(t_{n+1},y_{n+1}) = f_{n+1}$. Recall that $t_{n-1} = t_n - h$
     and $t_{n+1} = t_n + h$ and thus we have the four equations:

$$A(t_n-h)^2 + B(t_n-h) + C = y_{n-1} \qquad \text{(i)}$$
$$At_n^2 \quad\quad + Bt_n \quad\quad + C = y_n \qquad \text{(ii)}$$
$$A(t_n+h)^2 + B(t_n+h) + C = y_{n+1} \qquad \text{(iii)}$$
$$2A(t_n+h) + B = f_{n+1} \qquad \text{(iv)}$$

Subtracting Eq. (i) from Eq. (ii) to get Eq. (v) (not
shown) and subtracting Eq. (ii) from Eq. (iii) to get Eq.
(vi) (not shown), then subtracting Eq. (v) from Eq. (vi)
yields $y_{n+1} - 2y_n + y_{n-1} = 2Ah^2$, which can be solved for
A. Thus $B = f_{n+1} - 2A(t_n+h)$ [from Eq. (iv)] and
$C = y_n - t_n f_{n+1} + At_n^2 + 2At_n h$ [from Eq. (ii)]. Using
these values for A, B and C in Eq. (iv) yields
$y_{n+1} = (1/3)(4y_n - y_{n-1} + 2hf_{n+1})$, which is Eq. (15).

Section 8.6, Page 453

2a. If $0 \le t \le 1$ then we know $0 \le t^2 \le 1$ and hence
     $e^y \le t^2 + e^y \le 1 + e^y$. Since each of these terms
     represents a slope, we may conclude that the solution of
     Eq.(i) is bounded above by the solution of Eq.(iii) and
     is bounded below by the solution of Eq.(iv).

2b. $\phi_1(t)$ and $\phi_2(t)$ can each be found by separation of
     variables. For $\phi_1(t)$ we have $\dfrac{1}{1+e^y}dy = dt$, or

     $\dfrac{e^{-y}}{e^{-y}+1}dy = dt$. Integrating both sides yields
     $-\ln(e^{-y}+1) = t + c$. Solving for y we find
     $y = \ln[1/(c_1 e^{-t}-1)]$. Setting $t = 0$ and $y = 0$, we obtain
     $c_1 = 2$ and thus $\phi_1(t) = \ln[e^t/(2-e^t)]$. As $t \to \ln 2$, we
     see that $\phi_1(t) \to \infty$. A similar analysis shows that
     $\phi_2(t) = \ln[1/(c_2-t)]$, where $c_2 = 1$ when the I.C. are
     used. Thus $\phi_2(t) \to \infty$ as $t \to 1$ and thus we conclude
     that $\phi(t) \to \infty$ for some t such that $\ln2 \le t \le 1$.

2c. From Part b: $\phi_1(.9) = \ln[1/(c_1 e^{-.9}-1)] = 3.4298$ yields
     $c_1 = 2.5393$ and thus $\phi_1(t) \to \infty$ when $t \cong .9319$.
     Similarly for $\phi_2(t)$ we have $c_2 = .9324$ and thus
     $\phi_2(t) \to \infty$ when $t \cong .932$

3a. The general solution of the D.E. is $y(t) = t + ce^{\lambda t}$, where $y(0) = 0 \rightarrow c = 0$ and thus $y(t) = t$, which is independent of λ.

3c. Your result in Part b will depend upon the particular computer hardware and software that you use. If there is sufficient accuracy, you will obtain the solution $y = t$ for t on $0 \leq t \leq 1$ for each value of λ that is given, since there is no discretization error. If there is not sufficient accuracy, then round-off error will affect your calculations. For the larger values of λ, the numerical solution will quickly diverge from the exact solution, $y = t$, to the general solution $y = t + ce^{\lambda t}$, where the value of c depends upon the round-off error. If the latter case does not occur, you may simulate it by computing the numerical solution to the I.V.P. $y' - \lambda y = 1 - \lambda t$, $y(.1) = .10000001$. Here we have assumed that the numerical solution is exact up to the point $t = .09$ [i.e. $y(.09) = .09$] and that $t = .1$ round-off error has occurred as indicated by the slight error in the I.C. It has also been found that a larger step size ($h = .05$ or $h = .1$) may also lead to round-off error.

Section 8.7, Page 456

2a. The Euler formula is $\mathbf{x}_{n+1} = \mathbf{x}_n + h\mathbf{f}_n$, where

$$\mathbf{f}_n = \begin{pmatrix} 2x_n + t_n y_n \\ x_n y_n \end{pmatrix}.$$

Thus $\mathbf{f}_0 = \begin{pmatrix} 2-0 \\ (1)(1) \end{pmatrix} = \begin{pmatrix} 2 \\ 0 \end{pmatrix}$, $\mathbf{x}_1 = \begin{pmatrix} 1+.1(2) \\ 1+.1(1) \end{pmatrix} = \begin{pmatrix} 1.2 \\ 1.1 \end{pmatrix}$,

$\mathbf{f}_1 = \begin{pmatrix} 2.4+.1(1.1) \\ (1.2)(1.1) \end{pmatrix} = \begin{pmatrix} 2.51 \\ 1.32 \end{pmatrix}$

and $\mathbf{x}_2 \begin{pmatrix} 1.2+.1(2.51) \\ 1.1+.1(1.32) \end{pmatrix} = \begin{pmatrix} 1.451 \\ 1.232 \end{pmatrix} \cong \begin{pmatrix} \phi(.2) \\ \psi(.2) \end{pmatrix}$

2b. Eqs. (7) give:

$$\mathbf{k}_{01} = \begin{pmatrix} f(1,0) \\ g(1,0) \end{pmatrix} = \begin{pmatrix} 2-1 \\ (1)(1) \end{pmatrix} = \begin{pmatrix} 2 \\ 1 \end{pmatrix}$$

$$\mathbf{k}_{02} = \begin{pmatrix} 2.4+.1(1.1) \\ (1.2)(1.1) \end{pmatrix} = \begin{pmatrix} 2.51 \\ 1.32 \end{pmatrix}$$

$$\mathbf{k}_{03} = \begin{pmatrix} 2.502+.1(1.132) \\ (1.251)(1.132) \end{pmatrix} = \begin{pmatrix} 2.6152 \\ 1.41613 \end{pmatrix}$$

$$\mathbf{k}_{04} = \begin{pmatrix} 3.04608+.2(1.28323) \\ (1.52304)(1.28323) \end{pmatrix} = \begin{pmatrix} 3.30273 \\ 1.95440 \end{pmatrix}$$

Using Eq. (6) in scalar form, we then have

$x_1 = 1+(.2/6)[2+2(2.51)+2(2.6152)+3.30273] = 1.51844$

$y_1 = 1+(.2/6)[1+2(1.32)+2(1.41613)+1.95440] = 1.28089.$

7.   Write a computer program to do this problem as there are twenty steps or more for h ≤ .05.

8.   If we let $y = x'$, then $y' = x''$ and thus we obtain the system $x' = y$ and $y' = t-3x-t^2y$, with $x(0) = 1$ and $y(0) = x'(0) = 2$.   Thus $f(t,x,y) = y$, $g(t,x,y) = t - 3x - t^2y$, $t_0 = 0$, $x_0 = 1$ and $y_0 = 2$.   If a program has been written for an earlier problem, then its best to use that.   Otherwise, the first two steps are as follows:

$$\mathbf{k}_{01} = \begin{pmatrix} 2 \\ -3 \end{pmatrix}$$

$$\mathbf{k}_{02} = \begin{pmatrix} 2+(-.15) \\ .05-3(1.1)-(.05)^2(1.85) \end{pmatrix} = \begin{pmatrix} 1.85 \\ -3.25463 \end{pmatrix}$$

$$\mathbf{k}_{03} = \begin{pmatrix} 2+(-.16273) \\ .05-3(1.0925)-(.05)^2(1.83727) \end{pmatrix} = \begin{pmatrix} 1.83727 \\ -3.23209 \end{pmatrix}$$

$$\mathbf{k}_{04} = \begin{pmatrix} 2+(-.32321) \\ .1-3(1.18373)-(.1)^2(1.67679) \end{pmatrix} = \begin{pmatrix} 1.67679 \\ -3.46796 \end{pmatrix}$$

and thus

$x_1 = 1+(.1/6)[2 + 2(1.85)+2(1.83727)+(1.67679)]=1.18419,$

$y_1 = 2+(.1/6)[-3-2(3.25463)-2(3.23209)-3.46796]=1.67598,$

which are approximations to $x(.1)$ and $y(.1) = x'(.1).$

In a similar fashion we find

$$\mathbf{k}_{11} = \begin{pmatrix} 1.67598 \\ -3.46933 \end{pmatrix} \qquad\qquad \mathbf{k}_{12} = \begin{pmatrix} 1.50251 \\ -3.68777 \end{pmatrix}$$

$$\mathbf{k}_{13} = \begin{pmatrix} 1.49159 \\ -3.66151 \end{pmatrix} \qquad\qquad \mathbf{k}_{14} = \begin{pmatrix} 1.30983 \\ -3.85244 \end{pmatrix}$$

and thus

$x_2 = x_1 + (.1/6)[1.67598 + 2(1.50251) + 2(1.49159) + 1.30983] = 1.33376$
$y_2 = y_1 - (.1/6)[3.46933 + 2(3.68777) + 2(3.66151) + 3.85244] = 1.30897$.

Three more steps must be taken in order to approximate $x(.5)$ and $y(.5) = x'(.5)$. The intermediate steps yield $x(.3) \cong 1.44489$, $y(.3) \cong .9093062$ and $x(.4) \cong 1.51499$, $y(.4) \cong .4908795$.

CHAPTER 9

<u>Section 9.1, Page 469</u>

For Problems 1 through 16, once the eigenvalues have been
found, Table 9.1.1 will, for the most part, quickly yield the
type of critical point and the stability. In all cases it
can be easily verified that **A** is nonsingular.

1a. The eigenvalues are found from the equation $\det(\mathbf{A}-r\mathbf{I})=0$.
Substituting the values for **A** we have $\begin{vmatrix} 3-r & -2 \\ 2 & -2-r \end{vmatrix} =$
$r^2 - r + 2 = 0$ and thus the eigenvalues are $r_1 = -1$ and
$r_2 = 2$. For $r_1 = -1$, we have $\begin{pmatrix} 4 & -2 \\ 2 & -1 \end{pmatrix}\begin{pmatrix} \xi_1 \\ \xi_2 \end{pmatrix} = \begin{pmatrix} 0 \\ 0 \end{pmatrix}$ and thus
$\xi^{(1)} = \begin{pmatrix} 1 \\ 2 \end{pmatrix}$ and for r_2 we have $\begin{pmatrix} 1 & -2 \\ 2 & -4 \end{pmatrix}\begin{pmatrix} \xi_1 \\ \xi_2 \end{pmatrix} = \begin{pmatrix} 0 \\ 0 \end{pmatrix}$ and thus
$\xi^{(2)} = \begin{pmatrix} 2 \\ 1 \end{pmatrix}$.

1b. Since the eigenvalues differ in sign, the critical point
is a saddle point and is unstable.

1d.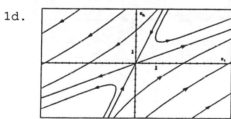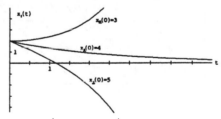

4a. Again the eigenvalues are given by $\begin{vmatrix} 1-r & -4 \\ 4 & -7-r \end{vmatrix} =$
$r^2 + 6r + 9 = 0$ and thus $r_1 = r_2 = -3$. The eigenvectors
are solutions of $\begin{pmatrix} 4 & -4 \\ 4 & -4 \end{pmatrix}\begin{pmatrix} \xi_1 \\ \xi_2 \end{pmatrix} = \begin{pmatrix} 0 \\ 0 \end{pmatrix}$ and hence there is
just one eigenvector $\xi = \begin{pmatrix} 1 \\ 1 \end{pmatrix}$.

4b. Since the eigenvalues are negative, $(0,0)$ is an improper
node which is asymptotically stable. If we had found
that there were two independent eigenvectors then $(0,0)$
would have been a proper node, as indicated in Case 3a.

4d.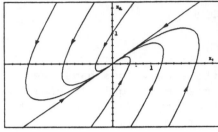

7a. In this case det$(A - rI) = r^2 - 2r + 5$ and thus the
     eigenvalues are $r_{1,2} = 1 \pm 2i$. For $r_1 = 1 + 2i$ we have

$$\begin{pmatrix} 2-2i & -2 \\ 4 & -2-2i \end{pmatrix} \begin{pmatrix} \xi_1 \\ \xi_2 \end{pmatrix} = \begin{pmatrix} 2-2i & -2 \\ 8-8i & -8 \end{pmatrix} \begin{pmatrix} \xi_1 \\ \xi_2 \end{pmatrix} = \begin{pmatrix} 0 \\ 0 \end{pmatrix} \text{ and thus}$$

$\xi^{(1)} = \begin{pmatrix} 1 \\ 1-i \end{pmatrix}$. Similarly for $r_2 = 1-2i$ we have

$$\begin{pmatrix} 2+2i & -2 \\ 4 & -2+2i \end{pmatrix} \begin{pmatrix} \xi_1 \\ \xi_2 \end{pmatrix} = \begin{pmatrix} 0 \\ 0 \end{pmatrix} \text{ and hence } \xi^{(2)} = \begin{pmatrix} 1 \\ 1+i \end{pmatrix}.$$

7b. Since the eigenvalues are complex with positive real
     part, we conclude that the critical point is a spiral
     point and is unstable.

7d.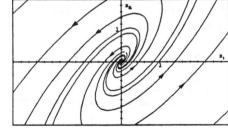

10a. Again, det$(A-rI) = r^2 + 9$ and thus we have $r_{1,2} = \pm 3i$.

     For $r_1 = 3i$ we have $\begin{pmatrix} 1-3i & 2 \\ -5 & -1-3i \end{pmatrix} \begin{pmatrix} \xi_1 \\ \xi_2 \end{pmatrix} = \begin{pmatrix} 0 \\ 0 \end{pmatrix}$ and thus

$\xi^{(1)} = \begin{pmatrix} 2 \\ 1-3i \end{pmatrix}$. Likewise for $r_2 = -3i$,

$$\begin{pmatrix} 1+3i & 2 \\ -5 & -1+3i \end{pmatrix} \begin{pmatrix} \xi_1 \\ \xi_2 \end{pmatrix} = \begin{pmatrix} 0 \\ 0 \end{pmatrix} \text{ so that } \xi^{(2)} = \begin{pmatrix} 2 \\ 1+3i \end{pmatrix}.$$

10b. Since the eigenvalues are pure imaginary the critical
     point is a center, which is stable.

10d.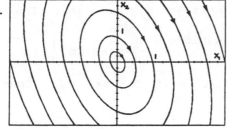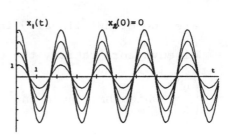

13. If we let $\mathbf{x} = \mathbf{x}^0 + \mathbf{u}$ then $\mathbf{x}' = \mathbf{u}'$ and thus the system becomes $\mathbf{u}' = \begin{pmatrix} 1 & 1 \\ 1 & -1 \end{pmatrix} \mathbf{x}^0 + \begin{pmatrix} 1 & 1 \\ 1 & -1 \end{pmatrix} \mathbf{u} - \begin{pmatrix} 2 \\ 0 \end{pmatrix}$ which will be in the form of Eq.(2) if $\begin{pmatrix} 1 & 1 \\ 1 & -1 \end{pmatrix} \mathbf{x}^0 = \begin{pmatrix} 2 \\ 0 \end{pmatrix}$. Using row operations, this last set of equations is equivalent to $\begin{pmatrix} 1 & 1 \\ 0 & -2 \end{pmatrix} \mathbf{x}^0 = \begin{pmatrix} 2 \\ -2 \end{pmatrix}$ and thus $x_1^0 = 1$ and $x_2^0 = 1$. Since $\mathbf{u}' = \begin{pmatrix} 1 & 1 \\ 1 & -1 \end{pmatrix} \mathbf{u}$ has $(0,0)$ as the critical point, we conclude that $(1,1)$ is the critical point of the original system. As in the earlier problems, the eigenvalues are given by $\begin{vmatrix} 1-r & 1 \\ 1 & -1-r \end{vmatrix} = r^2 - 2 = 0$ and thus $r_{1,2} = \pm\sqrt{2}$. Hence the critical point $(1,1)$ is an unstable saddle point.

17. The equivalent system is $dx/dt = y, dy/dt = -(k/m)x - (c/m)y$ which is written in the form of Eq.(2) as $\frac{d}{dt}\begin{pmatrix} x \\ y \end{pmatrix} = \begin{pmatrix} 0 & 1 \\ -k/m & -c/m \end{pmatrix} \begin{pmatrix} x \\ y \end{pmatrix}$. The point $(0,0)$ is clearly a critical point, and since $\underset{\sim}{A}$ is nonsingular, it is the only one. The characteristic equation is $r^2 + (c/m)r + k/m = 0$ so $r_1, r_2 = [-c \pm (c^2 - 4km)^{1/2}]/2m$. In the underdamped case $c^2 - 4km < 0$, the characteristic roots are complex with negative real parts and thus the critical point $(0,0)$ is an asymptotically stable spiral point. In the overdamped case $c^2 - 4km > 0$, the characteristic roots are real, unequal, and negative and hence the critical point $(0,0)$ is asymptotically stable node. In the critically damped case $c^2 - 4km = 0$, the characteristic roots are equal and negative. As indicated in the solution to Problem 4, to determine whether this is an improper or proper node we must determine whether there

are one or two linearly independent eigenvectors. The

eigenvectors satisfy the equations $\begin{pmatrix} c/2m & 1 \\ -k/m & -c/2m \end{pmatrix} \begin{pmatrix} \xi_1 \\ \xi_2 \end{pmatrix} =$

$\begin{pmatrix} 0 \\ 0 \end{pmatrix}$, which have just one solution if $c^2 - 4km = 0$. Thus

the critical point $(0,0)$ is an asymptotically stable
improper node.

18a. If **A** has one zero eigenvalue then for $r = 0$ we have
$\det(A-rI) = \det A = 0$. Hence **A** is singular which means
Ax = **0** has infinitely many solutions and consequently
there are infinitely many critical points.

18b. From Chapter 7, the solution is $x(t) = c_1\xi^{(1)} + c_2\xi^{(2)}e^{r_2 t}$,
which can be written in scalar form as
$x_1 = c_1\xi_1^{(1)} + c_2\xi_1^{(2)}e^{r_2 t}$ and $x_2 = c_1\xi_2^{(1)} + c_2\xi_2^{(2)}e^{r_2 t}$.
Assuming $\xi_1^{(2)} \neq 0$, the first equation can be solved for
$c_2 e^{r_2 t}$, which is then substituted into the second
equation to yield $x_2 = c_1\xi_2^{(1)} + [\xi_2^{(2)}/\xi_1^{(2)}][x_1-c_1\xi_1^{(1)}]$.
These are straight lines parallel to the vector $\xi^{(2)}$.
Note that the family of lines is independent of c_2. If
$\xi_1^{(2)} = 0$, then the lines are vertical. If $r_2 > 0$, the
direction of motion will be in the same direction as
indicated for $\xi^{(2)}$. If $r_2 < 0$, then it will be in the
opposite direction.

19b. Eq.(i) can be written in scalar form as $dx/dt = a_{11}x +$
$a_{12}y$ and $dy/dt = a_{21}x + a_{22}y$, which then yields Eq.(iii).
Ignoring the middle quotient in Eq.(iii), we can rewrite
that equation as $(a_{21}x + a_{22}y)dx - (a_{11}x + a_{12}y)dy = 0$,
which is exact since $a_{22} = -a_{11}$ from Eq.(ii)..

19c. Integrating $\phi_x = a_{21}x + a_{22}y$ we obtain $\phi = a_{21}x^2/2 + a_{22}xy$
$+ g(y)$ and thus $a_{22}x + g' = -a_{11}x - a_{12}y$ or $g' = -a_{12}y$
using Eq.(ii). Hence $a_{21}x^2/2 + a_{22}xy - a_{12}y^2/2 = k/2$ is
the solution to Eq.(iii). The quadratic equation $Ax^2 +$
$Bxy + Cy^2 = 0$ is an ellipse provided $B^2 - 4AC < 0$. Hence
for our problem if
$4a_{22}^2 + 4a_{21}a_{12} < 0$ then Eq.(iv) is an ellipse. Using
$a_{11} + a_{22} = 0$ we have $a_{22}^2 = -a_{11}a_{22}$ and hence

$-a_{11}a_{22} + a_{21}a_{12} < 0$ or $a_{11}a_{22} - a_{21}a_{12} > 0$, which is true by Eqs.(ii). Thus Eq.(iv) is an ellipse under the conditions of Eqs.(ii).

20. The given system can be written as $\dfrac{d}{dt}\begin{pmatrix} x \\ y \end{pmatrix} = \begin{pmatrix} a_{11} & a_{12} \\ a_{21} & a_{22} \end{pmatrix}\begin{pmatrix} x \\ y \end{pmatrix}$.

Thus the eigenvalues are given by
$r^2 - (a_{11}+a_{22})r + a_{11}a_{22} - a_{12}a_{21} = 0$ and using the given definitions we rewrite this as $r^2 - pr + q = 0$ and thus $r_{1,2} = (p \pm \sqrt{p^2-4q})/2 = (p \pm \sqrt{\Delta})/2$. The results are now obtained using Table 9.1.1.

Section 9.2, Page 478

1. Solutions of the D.E. for x are y are $x = Ae^{-t}$ and $y = Be^{-2t}$ respectively. $x(0) = 4$ and $y(0) = 2$ yield $A = 4$ and $B = 2$, so $x = 4e^{-t}$, and $y = 2e^{-2t}$. Solving the first equation for e^{-t} and

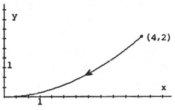

    then substituting into the second yields $y = 2[x/4]^2 = x^2/8$, which is a parabola. From the original D.E., or from the parametric solutions, we find that $0 < x \le 4$ and $0 < y \le 2$ for $t \ge 0$ and thus only the portion of the parabola shown is the trajectory, with the direction of motion indicated.

3. Utilizing the approach indicated in Eq.(11), we have $dy/dx = -x/y$, which separates into $xdx + ydy = 0$. Integration then yields the circle $x^2 + y^2 = c^2$, where $c^2 = 16$ for both sets of I.C. The direction of motion can be found from the original D.E. and is counterclockwise for both I.C. To obtain the parametric equations, we write the system in the form
    $\dfrac{d}{dt}\begin{pmatrix} x \\ y \end{pmatrix} = \begin{pmatrix} 0 & -1 \\ 1 & 0 \end{pmatrix}\begin{pmatrix} x \\ y \end{pmatrix}$, which has the characteristic
    equation $\begin{vmatrix} -r & -1 \\ 1 & -r \end{vmatrix} = r^2 + 1 = 0$, or $r = \pm i$. Following
    the procedures of Section 7.6, we find that one solution

of the above system is $\begin{pmatrix} 1 \\ -i \end{pmatrix} e^{it} = \begin{pmatrix} \cos t + i\sin t \\ \sin t - i\cos t \end{pmatrix}$ and thus

two real solutions are $\mathbf{u}(t) = \begin{pmatrix} \cos t \\ \sin t \end{pmatrix}$ and $\mathbf{v}(t) = \begin{pmatrix} \sin t \\ -\cos t \end{pmatrix}$.

The general solution of the system is then

$\begin{pmatrix} x \\ y \end{pmatrix} = c_1\mathbf{u}(t) + c_2\mathbf{v}(t)$ and hence the first I.C. yields

$c_1 = 4$, $c_2 = 0$, or $x = 4\cos t$, $y = 4\sin t$. The second I.C.
yields $c_1 = 0$, $c_2 = -4$, or $x = -4\sin t$, $y = 4\cos t$. Note
that both these parametric representations satsify the
form of the trajectories found in the first part of this
problem.

7a. The critical points are given by the solutions of
     $x(1-x-y) = 0$ and $y(1/2 - y/4 - 3x/4) = 0$. The solutions
     corresponding to either $x = 0$ or $y = 0$ are seen to be
     $x = 0$, $y = 0$; $x = 0$, $y = 2$; $x = 1$, $y = 0$. In addition,
     there is a solution corresponding to the intersection of
     the lines $1 - x - y = 0$ and $1/2 - y/4 - 3x/4 = 0$ which is
     the point $x = 1/2$, $y = 1/2$. Thus the critical points are
     $(0,0)$, $(0,2)$, $(1,0)$, and $(1/2,1/2)$.

7b.

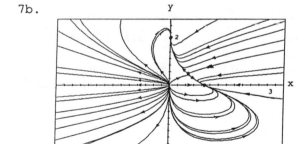

7c. For $(0,0)$ since all trajectories leave this point, this
     is an unstable node. For $(0,2)$ and $(1,0)$ since the
     trajectories tend to these points, respectively, they are
     asymptotically stable nodes. For $(1/2,1/2)$, one
     trajectory tends to $(1/2,1/2)$ while all others tend to
     infinity, so this is an unstable saddle point.

12a. The critical points are given by $y = 0$ and
     $x(1 - x^2/6 - y/5) = 0$, so $(0,0)$, $(\sqrt{6},0)$ and $(-\sqrt{6},0)$
     are the only critical points.

12b.

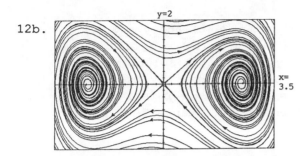

12c. Clearly $(\sqrt{6},0)$ and $(-\sqrt{6},0)$ are spiral points, and are asymptotically stable since the trajectories tend to each point, respectively. $(0,0)$ is a saddle point, which is unstable, since the trajectories behave like the ones for $(1/2,1/2)$ in Problem 7.

15. We know that $\phi'(t) = F[\phi(t), \psi(t)]$ and
$\psi'(t) = G[\phi(t), \psi(t)]$ for $\alpha < t < \beta$. By direct substitution we have
$\Phi'(t) = \phi'(t-s) = F[\phi(t-s), \psi(t-s)] = F[\Phi(t), \Psi(t)]$ and
$\Psi'(t) = \psi'(t-s) = G[\phi(t-s), \psi(t-s)] = G[\Phi(t), \Psi(t)]$ for
$\alpha < t-s < \beta$ or $\alpha+s < t < \beta+s$.

17. If $F(x_0,y_0,t) = G(x_0,y_0,t) = 0$ for all t then the conclusion is obvious. In a region where $F(x,y,t) \neq 0$ we may write

$$\frac{dy}{dx} = \frac{dy/dt}{dx/dt} = \frac{G(x,y,t)}{F(x,y,t)} = \frac{G(x,y,t)/[F^2(x,y,t) + G^2(x,y,t)]^{1/2}}{F(x,y,t)/[F^2(x,y,t) + G^2(x,y,t)]^{1/2}}$$

$$= \frac{A(x,y)}{B(x,y)}.$$

This is an autonomous D.E. which yields a one-parameter family of solutions independent of s. This family is the set of trajectories of the original system. A similar argument holds near points where $F(x,y,t) = 0$ but $G(x,y,t) \neq 0$ if we consider $dx/dy = F(x,y,t)/G(x,y,t)$.

18. Letting $x = \theta$ and $y = dx/dt$, then the D.E. can be written as the system $dx/dt = y$, $dy/dt = -\omega^2 \sin x$. Using the approach of Eq.(11), we find that this system has
$dy/dx = -\omega^2 \sin(x/y)$, which separates into
$y\,dy = -\omega^2 \sin x\,dx$. Integrating both sides yields
$y^2/2 = \omega^2 \cos x + c_1$. Setting $c_1 = c - \omega^2$ then yields the desired form.

21. Suppose that $t_1 > t_0$. Let $s = t_1 - t_0$. Since the system is autonomous, the result of Problem 15, with s replaced by $-s$ shows that $x = \phi_1(t+s)$ and $y = \psi_1(t+s)$ generates

the same trajectory (C_1) as $x = \phi_1(t)$ and $y = \psi_1(t)$.   But
at $t = t_0$ we have $x = \phi_1(t_0+s) = \phi_1(t_1) = x_0$ and
$y = \psi_1(t_0+s) = \psi_1(t_1) = y_0$. Thus the solution
$x = \phi_1(t+s)$, $y = \psi_1(t+s)$ satisfies <u>exactly</u> the same
initial conditions as the solution $x = \phi_0(t)$, $y = \psi_0(t)$
which generates the trajectory C_0.   Hence C_0 and C_1 are
the same.

22. From the existence and uniqueness theorem we know that if
    the two solutions $x = \phi(t)$, $y = \psi(t)$ and $x = x_0$, $y = y_0$
    satisfy $\phi(a) = x_0$, $\psi(a) = y_0$ and $x = x_0$, $y = y_0$ at $t = a$,
    then these solutions are identical. Hence $\phi(t) = x_0$ and
    $\psi(t) = y_0$ for all t contradicting the fact that the
    trajectory generated by $[\phi(t), \psi(t)]$ started at a
    noncritical point.

23. By direct substitution
    $\Phi'(t) = \phi'(t+T) = F[\phi(t+T), \psi(t+T)] = F[\Phi(t), \Psi(t)]$ and
    $\Psi'(t) = \psi'(t+T) = G[\phi(t+T), \psi(t+T)], G[\Phi(t), \Psi(t)]$.
    Furthermore $\Phi(t_0) = x_0$ and $\Psi(t_0) = y_0$. Thus by the
    existence and uniqueness theorem $\Phi(t) = \phi(t)$ and
    $\Psi(t) = \psi(t)$ for all t.

Section 9.3, Page 488

In Problems 1 through 4, write the system in the form of
Eq.(4).   Then if $g(0) = 0$ we may conclude that $(0,0)$ is a
critical point.   In addition, if g satisfies Eq.(5) or
Eq.(6), then the system is almost linear.   In this case the
linear system, Eq.(1), will determine, in most cases, the
type and stability of the critical point $(0,0)$ of the almost
linear system.   These results are summarized in Table 9.3.1.

3. In this case the system can be written as
    $$\frac{d}{dt}\begin{pmatrix} x \\ y \end{pmatrix} = \begin{pmatrix} 0 & 0 \\ -1 & 0 \end{pmatrix}\begin{pmatrix} x \\ y \end{pmatrix} + \begin{pmatrix} (1+x)\sin y \\ 1 - \cos y \end{pmatrix}.$$ However, the
    coefficient matrix is singular and $g_1(x,y) = (1+x)\sin y$
    does not satisfy Eq.(6). However, if we consider the
    Taylor series for $\sin y$, we see that $(1+x)\sin y - y =$
    $\sin y - y + x\sin y = -y^3/3! + y^5/5! + \cdots + x(y - y^3/3! + \cdots)$,
    which does satisfy Eq.(6), using $x = r\cos\theta$, $y = r\sin\theta$.
    Thus the first equation now becomes
    $$\frac{dx}{dt} = y + [(1+x)\sin y - y]$$ and hence

$$\frac{d}{dt}\begin{pmatrix} x \\ y \end{pmatrix} = \begin{pmatrix} 0 & 1 \\ -1 & 0 \end{pmatrix}\begin{pmatrix} x \\ y \end{pmatrix} + \begin{pmatrix} (1+x)\sin y - y \\ 1 - \cos y \end{pmatrix}, \text{ where the}$$

coefficient matrix is now nonsingular and

$$\mathbf{g}(x,y) = \begin{pmatrix} (1+x)\sin y - y \\ 1 - \cos y \end{pmatrix} \text{ satisfies Eq.(6).}$$

4. In this case the system can be written as

$$\frac{d}{dt}\begin{pmatrix} x \\ y \end{pmatrix} = \begin{pmatrix} 1 & 0 \\ 1 & 1 \end{pmatrix}\begin{pmatrix} x \\ y \end{pmatrix} + \begin{pmatrix} y^2 \\ 0 \end{pmatrix} \text{ and thus } \mathbf{A} = \begin{pmatrix} 1 & 0 \\ 1 & 1 \end{pmatrix} \text{ and}$$

$\mathbf{g} = \begin{pmatrix} y^2 \\ 0 \end{pmatrix}$. Since $\mathbf{g}(0) = \begin{pmatrix} 0 \\ 0 \end{pmatrix}$ we conclude that (0,0) is a

critical point. Following the procedure of Example 1, we
let $x = r\cos\theta$ and $y = r\sin\theta$ and thus

$$g_1(x,y)/r = r\frac{r^2 \sin^2\theta}{r} \to 0 \text{ as } r \to 0 \text{ and thus the system}$$

is almost linear. Since

$\det(\mathbf{A} - r\mathbf{I}) = (r-1)^2$, we find that the eigenvalues are
$r_1 = r_2 = 1$.   Since the roots are equal, we must

determine whether there are one or two eigenvectors to
classify the type of critical point. The eigenvectors

are determined by $\begin{pmatrix} 0 & 0 \\ 1 & 0 \end{pmatrix}\begin{pmatrix} \xi_1 \\ \xi_2 \end{pmatrix} = \begin{pmatrix} 0 \\ 0 \end{pmatrix}$ and hence there is

only one eigenvector $\xi = \begin{pmatrix} 0 \\ 1 \end{pmatrix}$. Thus the critical point

for the linear system is an unstable improper node. From
Table 9.3.1 we then conclude that the given system, which
is almost linear, has a critical point near (0,0) which
is either a node or spiral point (depending on how the
roots bifurcate) which is unstable.

6a. The critical points are the solutions of $x(1-x-y) = 0$ and
$y(3-x-2y) = 0$. Solutions are $x = 0$, $y = 0$; $x = 0$,
$3 - 2y = 0$ which give $y = 3/2$; $y = 0$ and $1 - x = 0$ which
give $x = 1$; and $1 - x - y = 0$, $3 - x - 2y = 0$ which give
$x = 1$, $y = 2$. Thus the critical points are (0,0),
(0,3/2), (1,0) and (-1,2).

6b, For the critical point (0,0) the D.E. is already in the
6c   form of an almost linear system; and the corresponding
     linear system is $du/dt = u$, $dv/dt = 3v$ which has the

eigenvalues $r_1 = 1$ and $r_2 = 3$. Thus the critical point (0,0) is an unstable node. Each of the other three critical points is dealt with in the same manner; we consider only the critical point (-1,2). In order to translate this critical point to the origin we set $x(t) = -1 + u(t)$, $y(t) = 2 + v(t)$ and substitute in the D.E. to obtain
$$du/dt = -1 + u - (-1+u)^2 - (-1+u)(2+v) = u + v - u^2 - uv$$
and
$$dv/dt = 3(2+v) - (-1+u)(2+v) - 2(2+v)^2 = -2u - 4v - uv - 2v^2.$$
Writing this in the form of Eq.(4) we find that
$$\mathbf{A} = \begin{pmatrix} 1 & 1 \\ -2 & -4 \end{pmatrix} \text{ and } \mathbf{g} = -\begin{pmatrix} u^2 + uv \\ uv + v^2 \end{pmatrix} \text{ which is an almost linear}$$
system. The eigenvalues of the corresponding linear system are $r = (-3 \pm \sqrt{9 + 8})/2$ and hence the critical point (-1,2), of the original system, is an unstable saddle point.

10a. The critical points are solutions of $x + x^2 + y^2 = 0$ and $y(1-x) = 0$, which yield (0,0) and (-1,0).

10b. For (0,0) the D.E. is already in the form of an almost linear system and thus $du/dt = u$ and $dv/dt = v$. For (-1,0) we let $u = x+1$, $v = y$ so that substituting $x = u-1$ and $y = v$ into the D.E. we obtain $\dfrac{du}{dt} = -u + u^2 + v^2$ and $\dfrac{dv}{dt} = 2v - uv$. Thus the corresponding linear system is $u' = -u$ and $v' = -2v$.

10c. For (0,0) $\mathbf{A} = \begin{pmatrix} 1 & 0 \\ 0 & 1 \end{pmatrix}$ which has $r_1 = r_2 = 1$, so that (0,0), for the non linear system, will be either a node or spiral point, depending on how the roots bifurcate. In any case, since r_1 and r_2 are positive, the system will be unstable. For (-1,0) $\mathbf{A} = \begin{pmatrix} -1 & 0 \\ 0 & 2 \end{pmatrix}$ and thus $r_1 = -1$ and $r_2 = 2$, and hence the nonlinear system, from Table 9.3.1, has an unstable saddle point at (-1,0).

18a. The system is $\dfrac{d}{dt}\begin{pmatrix} x \\ y \end{pmatrix} = \begin{pmatrix} 1 & 0 \\ 0 & -2 \end{pmatrix}\begin{pmatrix} x \\ y \end{pmatrix} + \begin{pmatrix} 0 \\ x^3 \end{pmatrix}$ and thus is almost linear using the procedures outlined in the

earlier problems. The corresponding linear system has
the eigenvalues $r_1 = 1$, $r_2 = -2$ and thus $(0,0)$ is an
unstable saddle point for both the linear and almost
linear systems.

18b. The trajectories of the linear system are the solutions
of $dx/dt = x$ and $dy/dt = -2y$ and thus $x(t) = c_1e^t$ and
$y(t) = c_2e^{-2t}$. To sketch these, solve the first equation
for e^t and substitute into the second to obtain
$y = c_1^2c_2/x^2$, $c_1 \neq 0$. Several
trajectories are shown in the
figure. Since $x(t) = c_1e^t$,
we must pick $c_1 = 0$ for
$x \to 0$ and $t \to \infty$. Thus $x = 0$,
$y = c_2e^{-2t}$ (the vertical axis)
is the only trajectory for
which $x \to 0$, $y \to 0$ as $t \to \infty$.

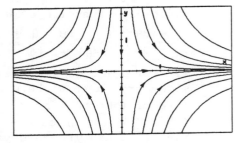

18c. For $x \neq 0$ we have $dy/dx = (dy/dt)/(dx/dt) = (-2y+x^3)/x$.
This is a linear equation, and the general solution is
$y = x^3/5 + k/x^2$, where k is an arbitrary constant. In
addition the system of equations has the solution $x = 0$,
$y = Be^{-2t}$. Any solution with its initial point on the
y-axis ($x=0$) is given by the latter solution. The
trajectories corresponding
to these solutions approach
the origin as $t \to \infty$. The
trajectory that passes through
the origin and divides the
family of curves is given by
$k = 0$, namely $y = x^3/5$. This
trajectory corresponds to the
trajectory $y = 0$ for the linear
problem. Several trajectories
are sketched in the figure.

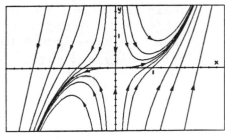

20. The equation of the trajectories was found in Problem 18
of Section 9.2.

22a.

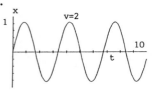
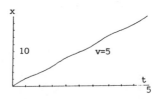
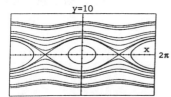

23a.

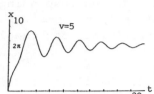 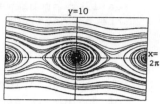

27a. Setting c = 0 in Eq.(17) of Section 9.2 and multiplying
by mL^2 we obtain $mL^2 d^2\theta/dt^2 + mgL\sin\theta = 0$. Considering
$d\theta/dt$ as a function of θ and using the chain rule we have
$$\frac{d}{dt}\left(\frac{d\theta}{dt}\right) = \frac{d}{d\theta}\left(\frac{d\theta}{dt}\right)\frac{d\theta}{dt} = \frac{1}{2}\frac{d}{d\theta}\left(\frac{d\theta}{dt}\right)^2 .$$ Thus
$(1/2)mL^2 d[(d\theta/dt)^2]/d\theta = -mgL\sin\theta$. Now integrate both
sides from α to θ where $d\theta/dt = 0$ at $\theta = \alpha$:
$(1/2)mL^2 (d\theta/dt)^2 = mgL(\cos\theta - \cos\alpha)$. Thus
$(d\theta/dt)^2 = (2g/L)(\cos\theta - \cos\alpha)$. Since we are releasing
the pendulum with zero velocity from a positive angle α,
the angle θ will initially be decreasing so $d\theta/dt < 0$.
If we restrict attention to the range of θ from $\theta = \alpha$ to
$\theta = 0$, we can assert $d\theta/dt = -\sqrt{2g/L}\,\sqrt{\cos\theta - \cos\alpha}$.
Solving for dt gives $dt = -\sqrt{L/2g}\,d\theta/\sqrt{\cos\theta - \cos\alpha}$.

27b. Since there is no damping, the pendulum will swing from
its initial angle α through 0 to $-\alpha$, then back through 0
again to the angle α in one period. It follows that
$\theta(T/4) = 0$. Integrating the last equation and noting
that as t goes from 0 to T/4, θ goes from α to 0 yields
$T/4 = -\sqrt{L/2g}\displaystyle\int_{\alpha}^{0} (1/\sqrt{\cos\theta - \cos\alpha})\,d\theta$.

28b. Under the given assumptions we have $g(x) = g(0) + g'(0)x$
$+ g''(\xi_1)x^2/2$ and $c(x) = c(0) + c'(\xi_2)x$, where
$0 < \xi_1,\ \xi_2 < x$ and $g(0) = 0$. Thus the system can be
written as
$$\frac{d}{dt}\begin{pmatrix} x \\ y \end{pmatrix} = \begin{pmatrix} 0 & 1 \\ -g'(0) & -c(0) \end{pmatrix}\begin{pmatrix} x \\ y \end{pmatrix} - \begin{pmatrix} 0 \\ -g''(\xi_1)x^2/2 - c'(\xi_2)xy \end{pmatrix},$$
from which the results follow.

Section 9.4, Page 502

3e.
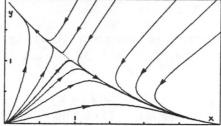

5b. The critical points are found by setting dx/dt = 0 and
     dy/dt = 0 and thus we need to solve x(1 - x - y) = 0 and
     y(1.5 - y - x) = 0. The first yields x = 0 or y = 1 - x
     and the second yields y = 0 or y = 1.5 - x. Thus (0,0),
     (0,3/2) and (1,0) are the only critical points since the
     two straight lines do not intersect in the first quadrant
     (or anywhere in this case). This is an example of one of
     the cases shown in Figure 9.4.5 a or b.

5e.
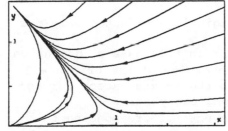

6b. The critical points are found by setting dx/dt = 0 and
     dy/dt = 0 and thus we need to solve x(1-x + y/2) = 0 and
     y(5/2 - 3y/2 + x/4) = 0. The first yields x = 0 or
     y = 2x - 2 and the second yields y = 0 or y = x/6 + 5/3.
     Thus we find the critical points (0,0), (1,0), (0,5/3)
     and (2,2). The last point is the intersection of the two
     straight lines, which will be used again in part d.

6c. For (0,0) the linearized system is $x' = x$ and $y' = 5y/2$,
     which has the eigenvalues $r_1 = 1$ and $r_2 = 5/2$. Thus the
     origin is an unstable node. For (2,2) we let
     x = u + 2 and y = v + 2 in the given system to find
     (since $x' = u'$ and $y' = v'$) that
     du/dt = (u+2)[1 - (u+2) + (v+2)/2] = (u+2)(-u+v/2) and
dv/dt = (v+2)[5/2 - 3(v+2)/2 + (u+2)/4] = (v+2)(u/4 - 3v/2).

     Hence the linearized equations are $\begin{pmatrix} u \\ v \end{pmatrix}' = \begin{pmatrix} -2 & 1 \\ 1/2 & -3 \end{pmatrix} \begin{pmatrix} u \\ v \end{pmatrix}$

     which has the eigenvalues $r_{1,2} = (-5 \pm \sqrt{3})/2$. Since
     these are both negative we conclude that (2,2) is an
     asymptotically stable node. In a similar fashion for

$(1,0)$ we let $x = u + 1$ and $y = v$ to obtain the linearized system $\begin{pmatrix} u \\ v \end{pmatrix}' = \begin{pmatrix} -1 & 1/2 \\ 0 & 11/4 \end{pmatrix} \begin{pmatrix} u \\ v \end{pmatrix}$. This has $r_1 = -1$ and $r_2 = 11/4$ as eigenvalues and thus $(1,0)$ is an unstable saddle point. Likewise, for $(0,5/3)$ we let $x = u$, $y = v + 5/3$ to find $\begin{pmatrix} u \\ v \end{pmatrix}' = \begin{pmatrix} 11/6 & 0 \\ 5/12 & -5/2 \end{pmatrix} \begin{pmatrix} u \\ v \end{pmatrix}$ as the corresponding linear system. Thus $r_1 = 11/6$ and $r_2 = -5/2$ and thus $(0,5/3)$ is an unstable saddle point.

6d. To sketch the required trajectories, we must find the eigenvectors for each of the linearized systems and then analyze the behavior of the linear solution near the critical point. Using this approach we find that the solution near $(0,0)$ has the form $\begin{pmatrix} x \\ y \end{pmatrix} = c_1 \begin{pmatrix} 1 \\ 0 \end{pmatrix} e^t + c_2 \begin{pmatrix} 0 \\ 1 \end{pmatrix} e^{5t/2}$ and thus the origin is approached only for large negative values of t. In this case e^t dominates $e^{5t/2}$ and hence in the neighborhood of the origin all trajectories are tangent to the x-axis except for one pair $(c_1 = 0)$ that lies along the y-axis.
For $(2,2)$ we find the eigenvector corresponding to $r = (-5 + \sqrt{3})/2 = -1.63$ is given by $(1-\sqrt{3})\xi_1/2 + \xi_2 = 0$ and thus $\begin{pmatrix} 1 \\ (\sqrt{3}-1)/2 \end{pmatrix} = \begin{pmatrix} 1 \\ .37 \end{pmatrix}$ is one eigenvector. For $r = (-5 - \sqrt{3})/2 = -3.37$ we have $(1 + \sqrt{3})\xi_1/2 + \xi_2 = 0$ and thus $\begin{pmatrix} 1 \\ -(\sqrt{3}+1)/2 \end{pmatrix} = \begin{pmatrix} 1 \\ -1.37 \end{pmatrix}$ is the second eigenvector.
Hence the linearized solution is $\begin{pmatrix} u \\ v \end{pmatrix} = c_1 \begin{pmatrix} 1 \\ .37 \end{pmatrix} e^{-1.63t} + c_2 \begin{pmatrix} 1 \\ -1.37 \end{pmatrix} e^{-3.37t}$. For large positive values of t the first term is the dominant one and thus we conclude that all trajectories but two approach $(2,2)$ tangent to the straight line with slope .37. If $c_1 = 0$, we see that there are exactly two $(c_2 > 0$ and $c_2 < 0)$ trajectories that lie on the straight line with slope -1.37.
In similar fashion, we find the linearized solutions near $(1,0)$ and $(0,5/3)$ to be, respectively,

$$\binom{u}{v} = c_1 \binom{1}{0} e^{-t} + c_2 \binom{1}{15/2} e^{11t/4} \text{ and}$$

$$\binom{u}{v} = c_1 \binom{0}{1} e^{-5t/2} + c_2 \binom{1}{5/52} e^{11t/6},$$

which, along with the
above analysis, yields
the sketch shown:

6e From the above sketch, it appears that $(x,y) \to (2,2)$ as
6f. $t \to \infty$ as long as (x,y) starts in the first quadrant.
To ascertain this, we need to prove that x and y cannot
become unbounded as $t \to \infty$. From the given system, we
can observe that, since $x > 0$ and $y > 0$, that dx/dt and
dy/dt have the same sign as the quantities $1 - x + y/2$
and $5/2 - 3y/2 + x/4$ respectively. If we set these
quantities equal to zero we get the straight lines
$y = 2x - 2$ and $y = x/6 + 5/3$, which divide the first
quadrant into the four
sectors shown. The signs
of x' and y' are indicated,
from which it can be
concluded that x and y
must remain bounded [and
in fact approach $(2,2)$] as
$t \to \infty$. The discussion
leading up to Fig.9.4.4
is also useful here.

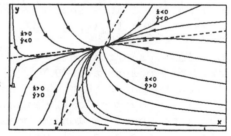

8a. Setting the right sides of the equations equal to zero
gives the critical points $(0,0)$, $(0, \varepsilon_2/\sigma_2)$, $(\varepsilon_1/\sigma_1, 0)$,
and possibly
$([\varepsilon_1\sigma_2 - \varepsilon_2\alpha_1]/[\sigma_1\sigma_2 - \alpha_1\alpha_2], [\varepsilon_2\sigma_1 - \varepsilon_1\alpha_2]/[\sigma_1\sigma_2 - \alpha_1\alpha_2])$.
(The last point can be obtained from Eq.(36) also). The
conditions $\varepsilon_2/\alpha_2 > \varepsilon_1/\sigma_1$ and $\varepsilon_2/\sigma_2 > \varepsilon_1/\alpha_1$ imply that
$\varepsilon_2\sigma_1 - \varepsilon_1\alpha_2 > 0$ and $\varepsilon_1\sigma_2 - \varepsilon_2\alpha_1 < 0$. Thus either the x
coordinate or the y coordinate of the last critical point
is negative so a mixed state is not possible. The
linearized system for $(0,0)$ is $x' = \varepsilon_1 x$ and $y' = \varepsilon_2 y$ and
thus $(0,0)$ is an unstable equilibrium point. Similarly,
it can be shown [by linearizing the given system or by
using Eq.(35)] that $(0, \varepsilon_2/\sigma_2)$ is an asymptotically
stable critical point and that $(\varepsilon_1\sigma_1, 0)$ is an unstable
critical point. Thus the fish represented by (redear)
survive.

8b. The conditions $\varepsilon_1/\sigma_1 > \varepsilon_2/\alpha_2$ and $\varepsilon_1/\alpha_1 > \varepsilon_2/\sigma_2$ imply that
     $\varepsilon_2\sigma_1 - \varepsilon_1\alpha_2 < 0$ and $\varepsilon_1\sigma_2 - \varepsilon_2\alpha_1 > 0$ so again one of the
     coordinates of the fourth point in 8a. is negative and
     hence a mixed state is not possible. An analysis similar
     to that in part(a) shows that $(0,0)$ and $(0,\varepsilon_2/\sigma_2)$ are
     unstable while $(\varepsilon_1/\sigma_1,0)$ is stable. Hence the bluegill
     (represented by x) survive in this case.

9b. If B is reduced, it is clear from the answer to part(a)
     that X is reduced and Y is increased. To determine
     whether the bluegill will die out, we give an intuitive
     argument which can be confirmed by doing the analysis.
     Note that $B/\gamma_1 = \varepsilon_1/\alpha_1 > \varepsilon_2/\sigma_2 = R$ and
     $R/\gamma_2 = \varepsilon_2/\alpha_2 > \varepsilon_1/\sigma_1 = B$ so that the graph of the lines
     $1 - x/B - \gamma_1 y/B = 0$ and $1 - y/R - \gamma_2 x/R = 0$ must appear
     as indicated in the figure,
     where critical points
     are inidcated by heavy
     dots. As B is decreased,
     X decreases, Y increases
     (as indicated above) and
     the point of intersection
     moves closer to $(0,R)$. If
     $B/\gamma_1 < R$ coexistence is not
     possible, and the only
     critical points are $(0,0),(0,R)$ and $(B,0)$.
     It can be shown that $(0,0)$ and $(B,0)$ are unstable and
     $(0,R)$ is asymptotically stable. Hence we conlcude, when
     coexistence is no longer possible, that $x \to 0$ and $y \to R$
     and thus the bluegill population will die out.

12c. Letting $x = u+2$ and $y = v+2$ yields
     $u' = (u+2)(4 - u-2 - v-2) = -2u - 2v - u^2 - uv$ and
     $v' = (v+2)(2 + 2\alpha - v-2 - \alpha u - 2\alpha) = -2\alpha u - 2v - v^2 - \alpha uv$.
     Thus the approximate linear system is $u' = -2u - 2v$ and
     $v' = -2\alpha u - 2v$.

12d. The eigenvalues are given by
$$\begin{vmatrix} -2-r & -2 \\ -2\alpha & -2-r \end{vmatrix} = r^2 + 4r + 4 - 4\alpha = 0, \text{ or } r = -2 \pm 2\sqrt{\alpha}.$$
     Thus for $0 < \alpha < 1$ there are 2 negative real roots
     (asymptotially stable node) and for $\alpha > 1$ the real roots
     differ in sign, yielding an unstable saddle point. $\alpha = 1$
     is the bifurcation point.

Section 9.5, Page 510

3b. We have x = 0 or (1 - .5x - .5y) = 0 and y = 0 or
(-.25 + .5x) = 0 and thus we have three critical points:
(0,0), (2,0) and (1/2,3/2).

3c. For (0,0) the linear system is dx/dt = x and

dy/dt = -.25y and hence A = $\begin{pmatrix} 1 & 0 \\ 0 & -1/4 \end{pmatrix}$ which has

eigenvalues r_1 = 1 and r_2 = -1/4 and corresponding

eigenvectors $\begin{pmatrix} 1 \\ 0 \end{pmatrix}$ and $\begin{pmatrix} 0 \\ 1 \end{pmatrix}$. Thus (0,0) is an unstable

saddle point.
For (2,0), we let x = 2 + u and y = v in the given

equations and obtain $\dfrac{du}{dt}$ = -(u+v) - $\dfrac{1}{2}$u(u+v) and

$\dfrac{dv}{dt}$ = $\dfrac{3}{4}$v + $\dfrac{1}{2}$uv. The linear portion of this has matrix

A = $\begin{pmatrix} -1 & -1 \\ 0 & 3/4 \end{pmatrix}$, which has the eigenvalues r_1 = -1,

r_2 = 3/4 and corresponding eigenvectors $\begin{pmatrix} 1 \\ 0 \end{pmatrix}$ and $\begin{pmatrix} -4 \\ 7 \end{pmatrix}$.

Thus (2,0) is also an unstable saddle point.

For $\begin{pmatrix} \dfrac{1}{2}, & \dfrac{3}{2} \end{pmatrix}$ we let x = 1/2 + u and y = 3/2 + v in the

given equations, which yields $\dfrac{du}{dt}$ = $-\dfrac{1}{4}$u - $\dfrac{1}{4}$v, $\dfrac{dv}{dt}$ = $\dfrac{3}{4}$u

as the linear portion. Thus A = $\begin{pmatrix} -\dfrac{1}{4} & -\dfrac{1}{4} \\ \dfrac{3}{4} & 0 \end{pmatrix}$, which has

eigenvalues $r_{1,2}$ = (-1 ± $\sqrt{11}$ i)/8. Thus $\begin{pmatrix} \dfrac{1}{2}, & \dfrac{3}{2} \end{pmatrix}$ is an

asymptotically stable spiral point since the eigenvalues
are complex with negative real part. Using
r_1 = (-1 + $\sqrt{11}$ i)/8 we find that one eigenvector is
$\begin{pmatrix} -2 \\ 1 + \sqrt{11}\,i \end{pmatrix}$ and by Section 7.6 the second eigenvector is
$\begin{pmatrix} -2 \\ 1 - \sqrt{11}\,i \end{pmatrix}$.

3e.

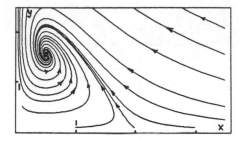

3f. For (x,y) above the line x + y = 2 we see that x' < 0 and
     thus x must remain bounded. For (x,y) to the right of
     x = 1/2, y' > 0 so it appears that y could grow large
     asymptotic to x = constant. However, this implies a
     contradiction (x = constant implies x' = 0, but as y gets
     larger, x' gets increasingly negative) and hence we
     conclude y must remain bounded and hence (x,y) →
     (1/2,3/2) as t → ∞, again assuming they start in the
     first quadrant.

7a. The amplitude ratio is $(cK/\gamma)/(\sqrt{ac}\,K/\alpha) = \alpha\sqrt{c}/\gamma\sqrt{a}$.

7b. In this case α = .5, a = 1, γ = .25 and c = .75, so the
     ratio is $.5\sqrt{.75}/.25\sqrt{1} = 2\sqrt{.75} = \sqrt{3}$.

7c. A rough measurement of the amplitudes is (6.1 - 1)/2 = 2.55
     and (3.8 -. 9)/2 = 1.45 and thus the ratio is approximately
     1.8. In this case the linear approximation is a good
     predictor.

10. The presence of a trapping company actually would require
     a modification of the equations, either by altering the
     coefficients or by including nonhomogeneous terms on the
     right sides of the D.E. The effects of indiscreminate
     trapping could decrease the populations of both rabbits
     and fox significantly or decrease the fox population
     which could possibly lead to a large increase in the
     rabbit population. Over the long run it makes sense for
     a trapping company to operate in such a way that a
     consistent supply of pelts is available and to distrub
     the predator-prey system as little as possible. Thus,
     the company should trap fox only when their population is
     increasing, trap rabbits only when their population is
     increasing, trap rabbits and fox only during the time
     when both their populations are increasing, and trap
     neither during the time both their populations are
     decreasing. In this way the trapping company can have a
     moderating effect on the population fluctuations, keeping
     the trajectory close to the center.

12. The critical points of the system are the solutions of the algebraic equations $x(a - \sigma x - \alpha y) = 0$, and $y(-c + \gamma x) = 0$. the critical points are $x = 0$, $y = 0$; $x = a/\sigma$, $y = 0$; and $x = c/\gamma$, $y = a/\alpha - c\sigma/\alpha\gamma = \sigma A/\alpha$ where $A = a/\sigma - c/\gamma > 0$.

To study the critical point $(0,0)$ we discard the nonlinear terms in the system of D.E. to obtain the corresponding linear system $dx/dt = ax$, $dy/dt = -cy$. The characteristic equation is $r^2 - (a+c)r - ac = 0$ so $r_1 = a$, $r_2 = -c$. Thus the critical point $(0,0)$ is an unstable saddle point.

To study the critical point $(a/\sigma, 0)$ we let $x = (a/\sigma) + u$, $y = 0 + v$ and substitute in the D.E. to obtain the almost linear system $du/dt = -au - (a\alpha/\sigma)v - \sigma u^2 - \alpha uv$, $dv/dt = \gamma Av + \gamma uv$. The corresponding linear system is $du/dt = -au - (a\alpha/\sigma)v$, $dv/dt = \gamma Av$. The characteristic equation is $r^2 + (a - \gamma A)r - a\gamma A = 0$ so $r_1 = -a$, $r_2 = \gamma A$. Thus the critical point $(a/\sigma, 0)$ is an unstable saddle point.

To study the critical point $(c/\gamma, \sigma A/\alpha)$ we let $x = (c/\gamma) + u$, $y = (\sigma A/\alpha) + v$ and substitute in the D.E. to obtain the almost linear system

$$\frac{du}{dt} = -(c\sigma/\gamma)u - (ac/\gamma)v - \sigma u^2 - \alpha uv$$

$$\frac{dv}{dt} = (\sigma A\gamma/\alpha)u + \gamma uv$$

The corresponding linear system is $du/dt = -(c\sigma/\gamma)u - (\alpha c/\gamma)v$, $dv/dt = (\sigma A\gamma/\alpha)u$. The characteristic equation is $r^2 + (c\sigma/\gamma)r + c\sigma A = 0$, so $r_1, r_2 = [-(c\sigma/\gamma) \pm \sqrt{(c\sigma/\gamma)^2 - 4c\sigma A}]/2$. Thus, depending on the sign of the discriminant we have that $(c/\gamma, \sigma A/\alpha)$ is either an asymptotically stable spiral point or an asymptotically stable node. Thus for nonzero initial data $(x,y) \to (c/\gamma, \sigma A/\alpha)$ as $t \to \infty$.

Section 9.6, Page 520

1. Assuming that $V(x,y) = ax^2 + cy^2$ we find $V_x(x,y) = 2ax$, $V_y = 2cy$ and thus Eq. (7) yields $\dot{V}(x,y) = 2ax(-x^3 + xy^2) + 2cy(-2x^2y - y^3) = -[2ax^4 + 2(2c-a)x^2y^2 + 2cy^4]$. If we choose a and c to be any positive real numbers with $2c > a$, then $\dot{V}$ is a negative definite. Also, V is positive definite by Theorem 9.6.4. Thus by Theorem 9.6.1 the origin is an asymptotically stable critical point.

3.   Assuming the same form for $V(x,y)$ as in Problem 1, we have

$$\dot{V}(x,y) = 2ax(-x^3 + 2y^3) + 2cy(-2xy^2) = -2ax^4 + 4(a-c)xy^3.$$

If we choose $a = c > 0$, then $\dot{V}(x,y) = -2ax^4 \leq 0$ in any

neighborhood containing the origin and thus $\dot{V}$ is negative semidefinite and V is positive definite. Theorem 9.6.1 then concludes that the origin is a stable critical point. Note that the origin may still be asymptotically stable, however, the $V(x,y)$ used here is not sufficient to prove that.

6a.   The correct system is $dx/dt = y$ and $dy/dt = -g(x)$. Since $g(0) = 0$, we conclude that $(0,0)$ is a critical point.

6b.   From the given conditions, the graph of g must be positive for $0 < x < k$ and negative for $-k < x < 0$. Thus
if $0 < x < k$ then $\int_0^x g(s)\,ds > 0,$

if $-k < x < 0$ then $\int_0^x g(s)\,ds = -\int_x^0 g(s)\,dx > 0.$

Since $V(0,0) = 0$ it follows that $V(x,y) = y^2/2 + \int_0^x g(s)\,ds$

is positive definite for $-k < x < k$, $-\infty < y < \infty$. Next,

we consider $\dot{V}(x,y) = g(x)y + y[-g(x)] = 0$. Since $\dot{V}(x,y)$ is never positive, we may conclude that it is negative semidefinite and hence by Theorem 9.6.1 $(0,0)$ is at least a stable critical point.

7b.   V is positive definite by Theorem 9.6.4. Since $V_x(x,y) = 2x$, $V_y(x,y) = 2y$, we obtain

$$\dot{V}(x,y) = 2xy - 2y^2 - 2y\sin x = 2y[-y + (x - \sin x)].$$ If

$x < 0$, then $\dot{V}(x,y) < 0$ for all $y > 0$. If $x > 0$, choose y

so that $0 < y < x - \sin x$. Then $\dot{V}(x,y) > 0$. Hence V is not a Liapunov function.

7c.   Since $V(0,0) = 0$, $1 - \cos x > 0$ for $0 < |x| < 2\pi$ and $y^2 > 0$ for $y \neq 0$, it follows that $V(x,y)$ is positive definite in a neighborhood of the origin. Next $V_x(x,y) = \sin x$, $V_y(x,y) = y$, so

$$\dot{V}(x,y) = (\sin x)(y) + y(-y - \sin x) = -y^2.$$ Hence $\dot{V}$ is negative semidefinite and $(0,0)$ is a stable critical point by Theorem 9.6.1.

7d. $V(x,y) = (x+y)^2/2 + x^2 + y^2/2 = 3x^2/2 + xy + y^2$ is
positive definite by Theorem 9.6.4. Next
$V_x(x,y) = 3x + y$, $V_y(x,y) = x + 2y$ so

$\dot{V}(x,y) = (3x+y)y - (x+2y)(y+\sin x)$
$\qquad = 2xy - y^2 - (x+2y)\sin x$
$\qquad = 2xy - y^2 - (x+2y)(x - \alpha x^3/6)$
$\qquad = -x^2 - y^2 + \alpha(x+2y)x^3/6$
$\qquad = -r^2 + \alpha r^4 (\cos\theta + 2\sin\theta)(\cos^3\theta)/6 < -r^2 + r^4/2$
$\qquad = -r^2(1-r^2/2)$. Thus $\dot{V}$ is negative definite for

$r < \sqrt{2}$. From Theorem 9.6.1 it follows that the origin
is an asymptotically stable critical point.

8.   Let $x = u$ and $y = du/dt$ to obtain the system $dx/dt = y$
and $dy/dt = -c(x)y - g(x)$. Now consider
$V(x,y) = y^2/2 + \int_0^x g(s)\,ds$.

10b. Since $V_x(x,y) = 2Ax + By$, $V_y(x,y) = Bx + 2Cy$, we have

$\dot{V}(x,y) = (2Ax + By)(a_{11}x + a_{12}y) + (Bx + 2Cy)(a_{21}x + a_{22}y) =$
$(2Aa_{11} + Ba_{21})x^2 + [2(Aa_{12} + Ca_{21}) + B(a_{11}+a_{22})]xy +$
$(2Ca_{22} + Ba_{12})y^2$. We choose A, B, and C so that
$2Aa_{11} + Ba_{21} = -1$, $2(Aa_{12} + Ca_{21}) + B(a_{11}+a_{22}) = 0$, and
$2Ca_{22} + Ba_{12} = -1$. The first and third equations give us A
and C in terms of B, respectively. We substitute in the
second equation to find B and then calculate A and C. The
result is given in the text.

10c. Since $a_{11}a_{22} - a_{12}a_{21} > 0$ and $a_{11} + a_{22} < 0$, we see that
$\Delta < 0$ and so $A > 0$. Using the expressions for A, B, and
C found in part (b) we obtain
$(4AC-B^2)\Delta^2 = [a_{21}^2 + a_{22}^2 + (a_{11}a_{22}-a_{12}a_{21})][a_{11}^2 + a_{12}^2 + (a_{11}a_{22}-a_{12}a_{21})]$
$\qquad\qquad\qquad - (a_{12}a_{22}+a_{11}a_{21})^2$
$\quad = (a_{11}^2+a_{12}^2+a_{21}^2+a_{22}^2)(a_{11}a_{22}-a_{12}a_{21}) + (a_{11}^2+a_{12}^2)(a_{21}^2+a_{22}^2)$
$\qquad\qquad\qquad + (a_{11}a_{22}-a_{12}a_{21})^2 - (a_{12}a_{22}+a_{11}a_{21})^2$
$\quad = (a_{11}^2+a_{12}^2+a_{21}^2+a_{22}^2)(a_{11}a_{22}-a_{12}a_{21}) + 2(a_{11}a_{22}-a_{12}a_{21})^2$.
Since $a_{11}a_{22} - a_{12}a_{21} > 0$ it follows that $4AC - B^2 > 0$.

11b. Substituting $x = r\cos\theta$, $y = r\sin\theta$ we find that

$\dot{V}[x(r,\theta), y(r,\theta)] = -r^2 + r(2A\cos\theta+B\sin\theta)F_1[x(r,\theta), y(r,\theta)]$
$+ r(B\cos\theta + 2C\sin\theta)G_1[x(r,\theta), y(r,\theta)]$. Now we make use
of the facts that (1) there exists an M such that
$|2A| \le M$, $|B| \le M$, and $|2C| \le M$; and (2) given any
$\varepsilon > 0$ there exists a circle $r = R$ such that

$|F_1(x,y)| < \varepsilon r$ and $|G_1(x,y)| < \varepsilon r$ for $0 \leq r < R$. We have $|2A\cos\theta + B\sin\theta| \leq 2M$ and $|B\cos\theta + 2C\sin\theta| \leq 2M$. Hence

$$\dot{V}[x(r,\theta), y(r,\theta)] \leq -r^2 + 2Mr(\varepsilon r) + 2Mr(\varepsilon r) = -r^2(1 - 4M\varepsilon).$$

If we choose $\varepsilon = M/8$ we obtain $\dot{V}[x(r,\theta), y(r,\theta)] \leq -r^2/2$

for $0 \leq r < R$. Hence $\dot{V}$ is negative definite in $0 \leq r < R$ and from Problem 10c V is positive definite and thus V is a Liapunov function for the almost linear system.

Section 9.7, Page 531

1. $r = 1$, $\theta = t + t_0$ is a periodic solution. If $r < 1$, then $dr/dt > 0$, and the direction of motion on a trajectory is outward. If $r > 1$, then the direction of motion is inward. It follows that the periodic solution $r = 1$, $\theta = t + t_0$ is a stable limit cycle.

2. $r = 1$, $\theta = -t + t_0$ is a periodic solution. If $r < 1$, then $dr/dt > 0$, and the direction of motion on a trajectory is outward. If $r > 1$, the $dr/dt > 0$, and the direction of motion is still outward. It follows that the solution $r = 1$, $\theta = -t + t_0$ is a semistable limit cycle.

4. $r = 1$, $\theta = -t + t_0$ and $r = 2$, $\theta = -t + t_0$ are periodic solutions. If $r < 1$, then $dr/dt < 0$, and the direction of motion on a trajectory is inward. If $1 < r < 2$, then $dr/dt > 0$, and the direction of motion is outward. Similarly, if $r > 2$, the direction of motion is inward. It follows that the periodic solution $r = 1$, $\theta = -t + t_0$ is unstable and the periodic solution $r = 2$, $\theta = -t + t_0$ is a stable limit cycle.

7. Differentiating x and y with respect to t we find that $dx/dt = (dr/dt)\cos\theta - (r\sin\theta)d\theta/dt$ and $dy/dt = (dr/dt)\sin\theta + (r\cos\theta)d\theta/dt$. Hence
$$ydx/dt - xdy/dt = (r\sin\theta\cos\theta)dr/dt - (r^2\sin^2\theta)d\theta/dt -$$
$$(r\cos\theta\sin\theta)dr/dt - (r^2\cos\theta)d\theta/dt$$
$$= -r^2 d\theta/dt.$$

8a. Multiplying the first equation by x and the second by y and adding yields $xdx/dt + ydy/dt = (x^2+y^2)f(r)/r$, or $rdr/dt = rf(r)$ and thus $dr/dt = rf(r)$. To obtain an equation for θ multiply the first equation by y, the second by x and substract to obtain

$ydx/dt - xdy/dt = -x^2-y^2$, or $-r^2d\theta/dt = -r^2$, using the results of Problem 7. Thus $d\theta/dt = 1$. It follows that periodic solutions are given by $r = c$, $\theta = t + t_0$ where $f(c) = 0$. Since $\theta = t + t_0$ the motion is counterclockwise.

8b. First note that $f(r) = r(r-2)^2(r-3)(r-1)$. Thus $r = 1$, $\theta = t + t_0$; $r = 2$, $\theta = t + t_0$; and $r = 3$, $\theta = t + t_0$ are periodic solutions. If $r < 1$, then $dr/dt > 0$, and the direction of motion on a trajectory is outward. If $1 < r < 2$, then $dr/dt < 0$ and the direction of motion is inward. Thus the periodic solution $r = 1$, $\theta = t + t_0$ is a stable limit cycle. If $2 < r < 3$, then $dr/dt < 0$, and the direction of motion is inward. Thus the periodic solution $r = 2$, $\theta = t + t_0$ is a semistable limit cycle. If $r > 3$, then $dr/dt > 0$, and the direction of motion is outward. Thus the periodic solution $r = 3$, $\theta = t + t_0$ is unstable.

9. Setting $x = r\cos\theta$, $y = r\sin\theta$ and using the techniques of Problem 8 the equations transform to $dr/dt = r^2 - 2$, $d\theta/dt = -1$. This system has a periodic solution $r = \sqrt{2}$, $\theta = -t + t_0$. If $r < \sqrt{2}$, then $dr/dt < 0$, and the direction of motion along a trajectory is inward. If $r > \sqrt{2}$, then $dr/dt > 0$, and the direction of motion is outward. Thus the periodic solution $r = \sqrt{2}$, $\theta = -t + t_0$ is unstable.

11. If $F(x,y) = x+y+x^3-y^2$, $G(x,y) = -x+2y+x^2y+y^3/3$, then $F_x(x,y) + G_y(x,y) = 1+3x^2+2+x^2+y^2 = 3+4x^2+y^2$. Since the conditions of Theorem 9.7.2 are satisfied for all x and y, it follows that the system has no periodic nonconstant solution.

13. Since $x = \phi(t)$, $y = \psi(t)$ is a solution of Eqs.(15), we have $d\phi/dt = F[\phi(t),\psi(t)]$, $d\psi/dt = G[\phi(t),\psi(t)]$. Hence on the curve C, $F(x,y)dy - G(x,y)dx = \phi'(t)\psi'(t)dt - \psi'(t)\phi'(t)dt = 0$. It follows that the line integral around C is zero. However, if $F_x + G_y$ has the same sign throughout D, then the double integral cannot be zero. This gives a contradiction. Thus either the solution of Eqs.(15) is not periodic or if it is, it cannot lie entirely in D.

16a. Setting $x' = 0$ and solving for y yields $y = x^3/3 - x + k$.
Substituting this into $y' = 0$ then gives
$x + .8(x^3/3 - x + k) - .7 = 0$ Using an equation solver we
obtain $x = 1.1994$ for $k = 0$ and $x = .80485$ for $k = .5$. To
determine the type of critical points these are, we use
Eq.(13) of Section 9.3 to find the linear coefficient matrix
to be $\mathbf{A} = \begin{pmatrix} 3(1-x_c^2) & 3 \\ -1/3 & -.8/3 \end{pmatrix}$, where x_c is the critical point.
For $x_c = 1.1994$ we obtain complex conjugate eigenvalues with
a negative real part, and therefore $k = 0$ yields an
asymptotically stable spiral point. For $x_c = .80485$ the
eigenvalues are also complex conjugates, but with positive
real parts, so $k = .5$ yields an unstable spiral point.

16b. Letting $k = .1, .2, .3, .4$ in the cubic equation of part (a)
and finding the corresponding eigenvalues from the matrix in
part (a), we find that the real part of the eigenvalues
change sign between $k = .3$ and $k = .4$. Continuing to
iterate in this fashion we find that for $k = .3464$ that the
real part of the eigenvalue is $-.0002$ while for $k = .3465$
the real part is $.00005$, which indicates $k_0 = .3465$ is the
critical point for which the system changes from stable to
unstable.

Section 9.8, Page 539

1b. For $\lambda = \lambda_3 = (-11 + \alpha)/2$, where $\alpha = \sqrt{81+40r}$, we have
$$\begin{pmatrix} -10+(11-\alpha)/2 & 10 & 0 \\ r & -1+(11-\alpha)/2 & 0 \\ 0 & 0 & -8/3+(11-\alpha)/2 \end{pmatrix} \begin{pmatrix} \xi_1 \\ \xi_2 \\ \xi_3 \end{pmatrix} = \begin{pmatrix} 0 \\ 0 \\ 0 \end{pmatrix}.$$
The last line implies $\xi_3 = 0$ and multiplying the first
line by
$(-9+\alpha)/2$ we obtain $\begin{pmatrix} (81-\alpha^2)/4 & 10(-9+\alpha)/2 \\ r & (9-\alpha)/2 \end{pmatrix} \begin{pmatrix} \xi_1 \\ \xi_2 \end{pmatrix} = \begin{pmatrix} 0 \\ 0 \end{pmatrix}.$
Substituting $\alpha^2 = 81+40r$ we have
$$\begin{pmatrix} -10r & -10(9-\alpha)/2 \\ r & (9-\alpha)/2 \end{pmatrix} \begin{pmatrix} \xi_1 \\ \xi_2 \end{pmatrix} = \begin{pmatrix} 0 \\ 0 \end{pmatrix}.$$
Thus $\xi^{(3)} = \begin{pmatrix} 9 - \sqrt{81+40r} \\ -2r \\ 0 \end{pmatrix}$, which is proportional to the
answer given in the text.

2a. The calculations are somewhat simplified if you let
$x = \beta + u$, $\underline{y = \beta + v}$, and $z = (r-1)+w$, where
$\beta = \sqrt{8(r-1)/3}$. An alternate approach is to extend
Eq.(13) of Section 9.3, which yields the same result.

2b. Eq.(9) is found by evaluating $\begin{vmatrix} -10-\lambda & 10 & 0 \\ 1 & -1-\lambda & -\beta \\ \beta & \beta & -8/3-\lambda \end{vmatrix} = 0$.

3b. If r_1, r_2, r_3 are the three roots of a cubic polynomial,
then the polynomial can be factored as
$(x-r_1)(x-r_2)(x-r_3)$. Expanding this and equating to the
given polynomial we have $A = -(r_1+r_2+r_3)$,
$B = r_1r_2 + r_1r_3 + r_2r_3$ and $C = -r_1r_2r_3$. We are interested
in the case when the real part of the complex conjugate
roots changes sign. Thus let $r_2 = \alpha+i\beta$ and $r_3 = \alpha-i\beta$,
which yields
$A = -(r_1+2\alpha)$, $\beta = 2\alpha r_1 + \alpha^2 + \beta^2$ and $C = r_1(\alpha^2+\beta^2)$.
Hence, if $AB = C$, we have
$-(r_1+2\alpha)(2\alpha r_1+\alpha^2+\beta^2) = -r_1(\alpha^2+\beta^2)$ or
$-2\alpha[r_1^2 + 2\alpha r_1 + (\alpha^2+\beta^2)] = 0$ or
$-2\alpha[(r_1+\alpha)^2+\beta^2] = 0$. Since the square bracket
term is positive, we conclude that if $AB = C$, then
$\alpha = 0$. That is, the conjugate complex roots are pure

imaginary. Note that the converse is also true. That
is, if the conjugate complex roots are pure imaginary
then $AB = C$.

4. We have
$\dot{V} = 2x[\sigma(-x+y)] + 2\sigma y[rx-y-xz] + 2\sigma z[-bz+xy]$
$= -2\sigma x^2 + 2\sigma xy + 2\sigma rxy - 2\sigma y^2 - 2\sigma bz^2$
$= 2\sigma\{-[x^2-(r+1)xy+y^2]-bz^2\}$. For $r < 1$, the term in the
square brackets remains positive for all values of x and

y, and thus $\dot{V}$ is negative definite. Thus, by the
extension of Theorem 9.6.1 to three equations, we
conclude that the origin is an asymptotically stable
critical point.

5a. $V = rx^2 + \sigma y^2 + \sigma(z-2r)^2 = c > 0$ yields
$\dfrac{dv}{dt} = 2rx[\sigma(-x+y)] + 2\sigma y(rx-y-xz) + 2\sigma(z-2r)(-bz+xy)$. Thus

$\dot{V} = -2\sigma[rx^2 + y^2 + b(z^2 - 2rz)]$
$= -2\sigma[rx^2 + y^2 + b(z-r)^2 - br^2]$.

5b. From the proof of Theorem 9.6.1, we find that we need to

show that V, as found in part a, is always negative as it
crosses $V(x,y,z) = c$. (Actually, we need to use the
extension of Theorem 9.6.1 to three equations, but the
proof is very similar using the vector calculus
approach.)   From part a we see that

$\dot{V} < 0$ if $rx^2 + y^2 + b(z-r)^2 > br^2$, which holds if (x,y,z)

lies outside the ellipsoid $\dfrac{x^2}{br} + \dfrac{y^2}{br^2} + \dfrac{(z-r)^2}{r^2} = 1$. (i)

Thus we need to choose c such that $V = c$ lies outside
Eq.(i). Writing $V = c$ in the form of Eq.(i) we obtain

the ellipsoid $\dfrac{x^2}{c/r} + \dfrac{y^2}{c/\sigma} + \dfrac{(z-2r)^2}{c/\sigma} = 1$. (ii) Now let

$M = \max(\sqrt{br},\ r\sqrt{b},\ r)$, then the ellipsoid (i) is

contained inside the sphere S1: $\dfrac{x^2}{M^2} + \dfrac{y^2}{M^2} + \dfrac{(z-r)^2}{M^2} = 1$.

Let S2 be a sphere centered at $(0,0,2r)$ with radius

$M+r$: $\dfrac{x^2}{(M+r)^2} + \dfrac{y^2}{(M+r)^2} + \dfrac{(z-2r)}{(M+r^2)} = 1$, then S1 is contained

in S2. Thus, if we choose c, in Eq.(ii), such that

$\dfrac{c}{r} > (M+r)^2$ and $\dfrac{c}{\sigma} > (M+r)^2$, then $\dot{V} < 0$ as the trajectory

crosses $V(x,y,z) = c$. Note that this is a sufficient
condition and there may be many other "better" choices
using different techniques.

8b. Several cases are shown. Results may vary, particularly
     for r = 24, due to the closeness of r to $r_3 \cong 24.06$.

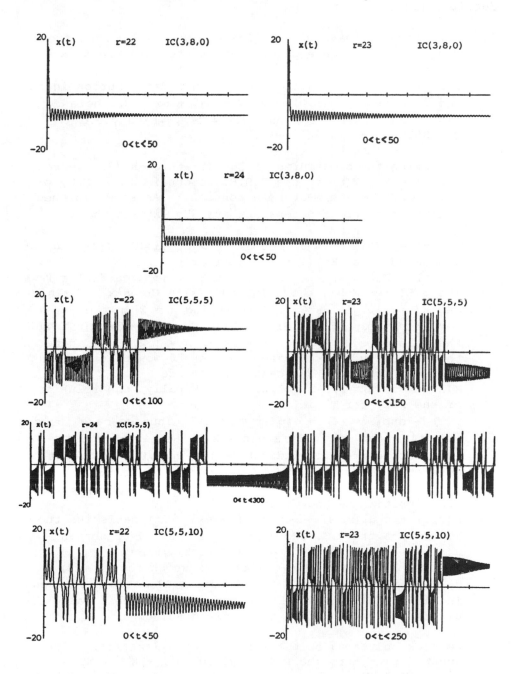

CHAPTER 10

Section 10.1, Page 550

2. For silver we have $\alpha^2 = 1.71\text{cm}^2/\text{sec} = .00171\text{m}^2/\text{sec}$ from
    Table 10.1.1 We are given that the temperature at $x = 0$
    is $30°C$ and at $x = 2\text{m}$ is $50°C$, so $u(0,t) = 30$ and
    $u(2,t) = 50$. The initial temperature distribution in the
    rod is given by $u(x,0) = f(x) = Ax^2 + Bx + C$, where
    $f(0) = 30°$, $f(1) = 60°$ and $f(2) = 50°$, which yield
    $C = 30$, $B = 50$ and $A = -20$.

5. Following the procedures of Eqs.(5) through (8), we set
    $u(x,y) = X(x)T(t)$ in the P.D.E. to obtain $X''T = 4XT'$, or
    $X''X = 4T'/T$, which must be a constant. As shown in the
    text this separation constant must be $-\lambda^2$ and thus
    $X'' + \lambda^2 X = 0$ and $T' + (\lambda^2/4)T = 0$. Now
    $u(0,t) = X(0)T(t) = 0$, for all $t > 0$, yields $X(0) = 0$, as
    discussed after Eq.(11) and similarly
    $u(2,t) = X(2)T(t) = 0$, for all $t > 0$, implies $X(2) = 0$.
    The D.E. for X has the solution $X(x) = C_1\cos\lambda x + C_2\sin\lambda x$
    and $X(0) = 0$ yields $C_1 = 0$. Setting $x = 2$ in the
    remaining form of X yields $X(2) = C_2\sin 2\lambda = 0$, which has
    the solutions $2\lambda = n\pi$ or $\lambda = n\pi/2$, $n = 1,2,\dots$. Note
    that we exclude $n = 0$ since then $\lambda = 0$ would yield
    $X(x) = 0$, which is unacceptable. Hence
    $X(x) = \sin(n\pi x/2)$, $n = 1,2,\dots$. Finally, the solution
    of the D.E. for T yields
    $T(t) = \exp(-\lambda^2 t/4) = \exp(-n^2\pi^2 t/16)$. Thus we have found
    $u_n(x,t) = \exp(-n^2\pi^2 t/16)\sin(n\pi x/2)$. Setting $t = 0$ in
    this last expression indicates that $u_n(x,0)$ has, for the
    correct choices of n, the same form as the terms in
    $u(x,0)$, the initial condition. Using the principle of
    superposition we know that
    $u(x,t) = c_1u_1(x,t) + c_2u_2(x,t) + c_4u_4(x,t)$ satisfies the
    P.D.E. and the B.C. and hence we let $t = 0$ to obtain
    $u(x,0) = c_1u_1(x,0) + c_2u_2(x,0) + c_4u_4(x,0) =$
    $c_1\sin\pi x/2 + c_2\sin\pi x + c_4\sin 2\pi x$. If we choose $c_1 = 2$,
    $c_2 = -1$ and $c_4 = 4$ then $u(x,0)$ here will match the given
    initial condition, and hence substituting these values in
    $u(x,t)$ above then gives the desired solution.

9. We seek solutions of the form $u(x,t) = X(x)T(t)$.
    Substituting into the P.D.E. yields $X''T + X'T' + XT' =$
    $X''T + (X' + X)T' = 0$. Formally dividing by the quantity
    $(X' + X)T$ gives the equation $X''/(X' + X) = -T'/T$ in which
    the variables are separated. In order for this equation

to be valid on the domain of u it is necessary that both
sides be equal to the same constant λ. Hence
$X''/(X' + X) = -T'/T = \lambda$ or equivalently, $X'' - \lambda(X' + X) = 0$
and $T' + \lambda T = 0$.

11. We seek solutions of the form $u(x,y) = X(x)Y(y)$.
Substituting into the P.D.E. yields $X''Y + (x+y)XY'' =$
$X''Y + xXY'' + yXY'' = 0$. Formally dividing by XY yields
$X''/X + xY''/Y + yY''/Y = 0$. From this equation we see that
the presence of the independent variable x multiplying
the term u_{yy} in the original equation leads to the term
xY''/Y when we attempt to separate the variables. It
follows that the argument for a separation constant does
not carry through and we cannot replace the P.D.E. by two
O.D.E.

14. Applying the chain rule to partial differentiation of u
with respect to x we see that $u_x = u_\xi \xi_x = u_\xi (1/L)$ and
$u_{xx} = u_{\xi\xi}(1/L)^2$. Substituting $u_{\xi\xi}/L^2$ for u_{xx} in the heat
equation gives $\alpha^2 u_{\xi\xi}/L^2 = u_t$ or $u_{\xi\xi}ß = (L^2/\alpha^2)u_t$. A
similar argument for the t substitution then yields the
desired P.D.E.

16. Substituting $u(x,y,t) = X(x)Y(y)T(t)$ in the P.D.E. yields
$\alpha^2(X''YT + XY''T) = XYT'$, which is equivalent to
$X''x + Y'' Y = T'/\alpha^2 T$. by keeping the independent
variablesx and y fixed and varying t we see that $T'/\alpha^2 T$
must equal some constant σ_1 since the left side of the
equation is fixed. Hence, $X''/X + Y''/Y = T'/\alpha^2 T = \sigma_1$, or
$X''/X = \sigma_1 - Y''/Y$ and $T' - \sigma_1\alpha^2 T = 0$. by keeping x fixed
and varying y in the equation involving X and Y we see
that $\sigma_1 - Y''/Y$ must equal some constant σ_2 since the left
side of the equation is fixed. Hence,
$X''/X = \sigma_1 - Y''/Y = \sigma_2$ so $X'' - \sigma_2 X = 0$ and
$Y'' - (\sigma_1 - \sigma_2)Y = 0$. If we assume homogeneous B.C. and
require that solutions be bounded as $t \to \infty$, then it can
be shown by an argument similar to the one in the text
that σ_1 and σ_2 must be negative constants,
$\sigma_1 = -\lambda^2$, $\sigma_2 = -\mu^2$. Then the O.D.E. become
$X'' + \mu^2 X = 0$, $Y'' + (\lambda^2 - \mu^2)Y = 0$, and $T' + \alpha^2\lambda^2 T = 0$.

Section 10.2, Page 560

3. We look for values of T for which $\sinh 2(x+T) = \sinh 2x$ for
all x. Expanding the left side of this equation gives

sinh2xcosh2T + cosh2xsinh2T = sinh2x, which will be
satisfied for all x if we can choose T so that cosh2T = 1
and sinh2T = 0. The only value of T satisfying these two
constraints is T = 0. Since T is not positive we
conclude that the function sinh2x is not periodic.

5.   We look for values of T for which $\tan\pi(x+T) = \tan\pi x$.
      Expanding the left side gives
      $\tan\pi(x+T) = (\tan\pi x + \tan\pi T)/(1-\tan\pi x\tan\pi T)$ which is equal
      to $\tan\pi x$ only for $\tan\pi T = 0$. The only positive solutions
      of this last equation are T = 1,2,3... and hence $\tan\pi x$ is
      periodic with fundamental period T = 1.

7.   It may be helpful to draw a graph of the given function.

10.   The graph of f(x) is:
      We note that f(x) is
      a straight line with
      a slope of one in any
      interval. Thus f(x) has the form x+b in any interval for
      the correct value of b. Since f(x+2) = f(x), we may set
      x = -1/2 to obtain f(3/2) = f(-1/2). Noting that 3/2 is
      on the interval 1 < x < 2[f(3/2) = 3/2 + b] and that -1/2
      is on the interval -1 < x < 0[f(-1/2) = -1/2 + 1], we
      conclude that 3/2 + b = -1/2 + 1, or b = -1 for the
      interval 1 < x < 2. For the interval 8 < x < 9 we have
      f(x+8) = f(x+6) = ... = f(x) by successive applications
      of the periodicity condition. Thus for x = 1/2 we have
      f(17/2) = f(1/2) or 17/2 + b = 1/2 so b = -8 on the
      interval 8 < x < 9.

In Problems 13 through 18 it is often necessary to use
integration by parts to evaluate the coefficients, although
all the details will not be shown here.

13a. The function represents
      a sawtooth wave. It is
      periodic with period 2L.

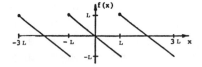

13b. The Fourier series is of the form

$$f(x) = a_0/2 + \sum_{m=1}^{\infty} (a_m\cos m\pi x/L + b_m\sin m\pi x/L)$$ where the

coefficients are computed from Eqs. (13) and (14).
Substituting for f(x) in these equations yields
$a_0 = (1/L)\int_{-L}^{L} (-x)\,dx = 0$; $a_m = (1/L)\int_{-L}^{L} (-x)\cos(m\pi x/L)\,dx = 0$,
m = 1,2...; and

$$b_m = (1/L) \int_{-L}^{L} (-x) \sin(m\pi x/L) \, dx$$

$$= (x/m\pi) \cos(m\pi x/L) \Big|_{-L}^{L} - (1/m\pi) \int_{-L}^{L} \cos(m\pi x/L) \, dx$$

$$= (2L\cos m\pi)/m\pi - (L/m^2\pi^2) \sin(m\pi x/L) \Big|_{-L}^{L} = 2L(-1)^m/m\pi$$

Substituting these terms in the above Fourier series for f(x) yields the desired answer.

15b. In this case f(x) is periodic of period 2π and thus $L = \pi$ in Eqs. (9), (13,) and (14). The constant a_0 is found to be $a_0 = (1/\pi) \int_{-\pi}^{0} x \, dx = -\pi/2$ since f(x) is zero on the interval $[0,\pi]$. Likewise $a_n = (1/\pi) \int_{-\pi}^{0} x \cos nx \, dx = [1 - (-1)^n] n^2\pi$ using integration by parts. Thus $a_n = 0$ for n even and $a_n = 2/n^2\pi$ for n odd, which may be written as $a_{2n-1} = 2/(2n-1)^2\pi$ since 2n-1 is always an odd number. In a similar fashion $b_n = (1/\pi) \int_{-\pi}^{0} x \sin nx \, dx = (-1)^{n+1}/n$ and thus the desired solution is obtained. Notice that in this case both cosine and sine terms appear in the Fourier series for the given f(x).

15a.   21a.

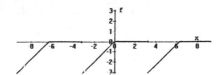
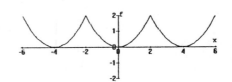

21b. $a_0 = \dfrac{1}{2} \int_{-2}^{2} \dfrac{x^2}{2} \, dx = \dfrac{1}{12} x^3 \Big|_{-2}^{2} = \dfrac{4}{3}$, so $\dfrac{a_0}{2} = \dfrac{2}{3}$ and

$a_n = \dfrac{1}{2} \int_{-2}^{2} \dfrac{x^2}{2} \cos \dfrac{n\pi x}{2} \, dx$

$= \dfrac{1}{4} [\dfrac{2x^2}{n\pi} \sin \dfrac{n\pi x}{2} + \dfrac{8x}{n^2\pi^2} \cos \dfrac{n\pi x}{2} - \dfrac{16}{n^3\pi^3} \sin \dfrac{n\pi x}{2}] \Big|_{-2}^{2}$

$= (8/n^2\pi^2) \cos(n\pi) = (-1)^n 8/n^2\pi^2$

where the second line for a_n is found by integration by parts or a computer algebra system. Similarly,

$$b_n = \frac{1}{2}\int_{-2}^{2}\frac{x^2}{2}\frac{n\pi x}{2}dx$$

$$= \frac{1}{4}[\frac{-2x^2}{n\pi}\cos\frac{n\pi x}{2} + \frac{8x}{n^2\pi^2}\sin\frac{n\pi x}{2} + \frac{16}{n^3\pi^3}\cos\frac{n\pi x}{2}]_{-2}^{2} = 0.$$

Thus $f(x) = \frac{2}{3} + \frac{8}{\pi^2}\sum_{n=1}^{\infty}\frac{(-1)^n}{n^2}\cos\frac{n\pi x}{2}$.

21c. As in Eq. (27), we have $s_m(x) = \frac{2}{3} + \frac{8}{\pi^2}\sum_{n=1}^{m}\frac{(-1)^n}{n^2}\cos\frac{n\pi x}{2}$

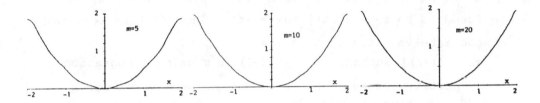

21d. Observing the graphs we see that the Fourier series converges quite rapidly.

25.

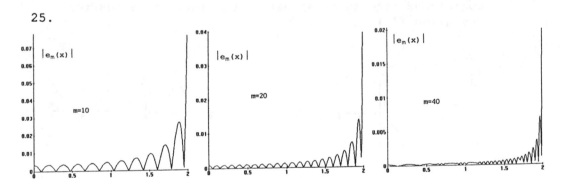

27a. First we have $\int_{T}^{a+T}g(x)dx = \int_{0}^{a}g(s)ds$ by letting $x = s + T$ in the left integral. Now, if $0 \le a \le T$, then from elementary calculus we know that $\int_{a}^{a+T}g(x)dx =$ $\int_{a}^{T}g(x)dx + \int_{T}^{a+T}g(x)dx = \int_{a}^{T}g(x)dx + \int_{0}^{a}g(x)dx$ using the equality derived above. This last sum is $\int_{0}^{T}g(x)dx$ and thus we have the desired result.

Section 10.3, Page 567

2a. Substituting for f(x) in Eqs.(2) and (3) with L = π
     yields $a_0 = (1/\pi) \int_0^\pi x dx = \pi/2$;

     $a_m = (1/\pi) \int_0^\pi x \cos mx dx = (\cos m\pi - 1)/\pi m^2 = 0$ for m even and

     $= -2/\pi m^2$ for m odd; and

     $b_m = (1/\pi) \int_0^\pi x \sin mx dx = -(\pi \cos m\pi)/m\pi = (-1)^{m+1}/m$,

     m = 1,2... . Substituting these values into Eq.(1) with
     L = π yields the desired solution.

2b. The function to which the series converges is indicated
     in the figure and is periodic with period 2π. Note that

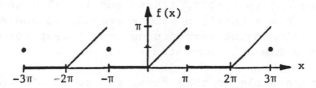

     the Fourier series converges to π/2 at x = −π, π, etc.,
     even though the function is defined to be zero there.
     This value (π/2) represents the mean value of the left
     and right hand limits at those points. In (−π, 0),
     f(x) = 0 and f'(x) = 0 so both f and f' are continuous and
     have finite limits as x → −π from the right and as
     x → 0 from the left. In (0, π), f(x) = x and f'(x) = 1
     and again both f and f' are continuous and have limits as
     x → 0 from the right and as x → π from the left. Since
     f and f' are piecewise continuous on [−π, π] the
     conditions of the Fourier theorem are satisfied.

4a. Substituting for f(x) in Eqs.(2) and (3), with L = 1
     yields $a_0 = \int_{-1}^1 (1-x^2) dx = 4/3$;

     $a_n = \int_{-1}^1 (1-x^2) \cos n\pi x dx = (2/n\pi) \int_{-1}^1 x \sin n\pi x dx =$

     $(-2/n^2\pi^2) [x \cos n\pi x \big|_{-1}^1 - \int_{-1}^1 \cos n\pi x dx] =$

     $4(-1)^{n+1}/n^2\pi^2$; $b_n = \int_{-1}^1 (1-x^2) \sin n\pi x dx = 0$. Substituting

     these values into Eq.(1) gives the desired series.

4b. The function to which the series converges is shown in
     the figure and is periodic of fundamental period 2. In
     [−1,1] f(x) and f'(x) = −2x are both continuous and have
     finite limits as the endpoints of the interval are
     approached from within the interval.

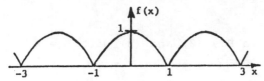

7a. As in Problem 15, Section 10.2, we have

$$f(x) = -\frac{\pi}{4} + \sum_{n=1}^{\infty}[\frac{2\cos(2n-1)x}{\pi(2n-1)^2} + \frac{(-1)^{n+1}\sin nx}{n}].$$

7b. $e_n(x) = f(x) + \frac{\pi}{4} - \sum_{k=1}^{n}[\frac{2\cos(2k-1)x}{\pi(2k-1)^2} + \frac{(-1)^{k+1}\sin kx}{k}].$

Using a computer algebra system, we find that for
$n = 5$, 10 and 20 the maximum error occurs at $x = -\pi$ in
each case and is 1.6025, 1.5867 and 1.5787 respectively.
Note that the author's n values are 10, 20 and 40, since
he has included the zero cosine coefficient terms and the
sine terms are all zero at $x = -\pi$.

7c. It's not possible in this case, due to Gibb's phenomenon,
to satisfy $|e_n(x)| \leq 0.01$ for all x.

12a. $b_n = \int_{-1}^{1}(x-x^3)\sin n\pi x\, dx$

$= [\frac{x^3}{n\pi}\cos n\pi x - \frac{3x^2}{n^2\pi^2}\sin n\pi x - \frac{(n^2\pi^2+6)}{n^3\pi^3}x\cos n\pi x + \frac{(n^2\pi^2+6)}{n^4\pi^4}\sin n\pi x]_{-1}^{1}$

$= \frac{-12}{n^3}\cos n\pi$, so $f(x) = -\frac{12}{\pi^3}\sum_{n=1}^{\infty}\frac{(-1)^n}{n^3}\sin n\pi x$.

12b. These errors will be much smaller than in the earlier
problems due to the n^3 factor in the denominator.
Convergence is much more rapid in this case.

14. The solution to the corresponding homogeneous equation is
found by methods presented in Section 3.4 and is
$y(t) = c_1\cos\omega t + c_2\sin\omega t$. For the nonhomogeneous terms
we use the method of superposition and consider the
sequence of equations $y_n'' + \omega^2 y_n = b_n\sin nt$ for
$n = 1,2,3\ldots$. If $\omega > 0$ is not equal to an integer,
then the particular solution to this last equation has
the form $Y_n = a_n\cos nt + d_n\sin nt$, as previously discussed
in Section 3.6. Substituting this form for Y_n into the
equation and solving, we find $a_n = 0$ and $d_n = b_n/(\omega^2-n^2)$.
Thus the formal general solution of the original
nonhomogeneous D.E. is

$y(t) = c_1 \cos\omega t + c_2 \sin\omega t + \sum\limits_{n=1}^{\infty} b_n(\sin nt)/(\omega^2-n^2)$, where

we have superimposed all the Y_n terms to obtain the infinite sum. To evalute c_1 and c_2 we set $t = 0$ to obtain $y(0) = c_1 = 0$ and

$y'(0) = \omega c_2 + \sum\limits_{n=1}^{\infty} n b_n/(\omega^2-n^2) = 0$ where we have formally

differentiated the infinite series term by term and evaluated it at $t = 0$. (Differentiation of a Fourier Series has not been justified yet and thus we can only consider this method a formal solution at this point).

Thus $c_2 = -(1/\omega) \sum\limits_{n=1}^{\infty} n b_n/(\omega^2-n^2)$, which when substituted

into the above series yields the desired solution.

If $\omega = m$, a positive integer, then the particular solution of $y_m'' + m^2 y_m = b_m \sin mt$ has the form $Y_m = t(a_m \cos mt + d_m \sin mt)$ since $\sin mt$ is a solution of the related homogeneous D.E. Substituting Y_m into the D.E. yields $a_m = -b_m/2m$ and $d_m = 0$ and thus the general solution of the D.E. (when $\omega = m$) is now $y(t) = c_1 \cos mt$

$+ c_2 \sin mt - b_m t(\cos mt)/2m + \sum\limits_{n=1,\,n\ne m}^{\infty} b_n(\sin nt)/(\omega^2-n^2)$. To

evaluate c_1 and c_2 we set $y(0) = c_1 = 0$ and

$y'(0) = c_2 m - b_m/2m + \sum\limits_{n=1,\,n\ne m}^{\infty} b_n n/(\omega^2-n^2) = 0$. Thus

$c_2 = b_m/2m^2 - \sum\limits_{n=1,\,n\ne m}^{\infty} b_n n/m(\omega^2-n^2)$, which when substituted

into the equation for $y(t)$ yields the desired solution.

15. In order to use the results of Problem 14, we must first find the Fourier series for the given $f(t)$. Using Eqs.(2) and (3) with $L = \pi$, we find that
$a_0 = (1/\pi)\int_0^\pi dx - (1/\pi)\int_\pi^{2\pi} dx = 0;$

$a_n = (1/\pi)\int_0^\pi \cos nx\,dx - (1/\pi)\int_\pi^{2\pi}\cos nx\,dx = 0;$ and

$b_n = (1/\pi)\int_0^\pi \sin nx\,dx - (1/\pi)\int_\pi^{2\pi}\sin nx\,dx = 0$ for n even and
$= 4/n\pi$ for n odd. Thus

$f(t) = (4/\pi) \sum_{n=1}^{\infty} \sin(2n-1)t/(2n-1)$. Comparing this to the forcing function of Problem 14 we see that b_n of Problem 14 has the specific values $b_{2n} = 0$ and $b_{2n-1} = (4/\pi)/(2n-1)$ in this example. Substituting these into the answer to Problem 14 yields the desired solution. Note that we have asumed ω is not a positive integer. Note also, that if the solution to Problem 14 is not available, the procedure for solving this problem would be exactly the same as shown in Problem 14.

16. In this case the Fourier series for $f(t)$ is given by

$f(t) = 1/2 + (4/\pi^2) \sum_{n=1}^{\infty} \cos(2n-1)\pi t/(2n-1)^2$ and thus we may

not use the form of the answer in Problem 14. The procedure outlined there, however, is applicable and will yield the desired solution.

18b. Note that f and f' are continuous at all points where f'' is continuous. Let $x_1, \ldots, x_m$ be the points in $(-L, L)$ where f'' is not continuous. By splitting up the interval of integration at these points, and integrating Eq.(3) by parts twice, we obtain

$$n^2 b_n = \frac{n}{\pi} \sum_{i=1}^{m} [f(x_i+) - f(x_i-)] \cos\frac{n\pi x_i}{L} - \frac{n}{\pi}[f(L-) - f(-L+)]\cos n\pi$$

$$-\frac{L}{\pi^2} \sum_{i=1}^{m} [f'(x_i+) - f'(x_1-)] \sin\frac{n\pi x_i}{L} - \frac{L}{\pi^2}\int_{-L}^{L} f''(x) \sin\frac{n\pi x}{L} dx, \text{ where}$$

we have used the fact that cosine is continuous. We want the first two terms on the right side to be zero, for otherwise they grow in magnitude with n. Hence f must be continuous throughout the closed interval $[-L, L]$. The last two terms are bounded, by the hypotheses on f' and f''. Hence $n^2 b_n$ is bounded; similarly $n^2 a_n$ is bounded. Convergence of the Fourier series then follows by

comparison with $\sum_{n=1}^{\infty} n^{-2}$.

Section 10.4, Page 575

3. Let $f(x) = \tan 2x$, then

$$f(-x) = \tan(-2x) = \frac{\sin(-2x)}{\cos(-2x)} = \frac{-\sin(2x)}{\cos(2x)} = -\tan 2x = -f(x)$$

and thus $\tan 2x$ is an odd function.

7.

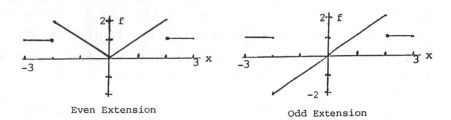

Even Extension   Odd Extension

10.

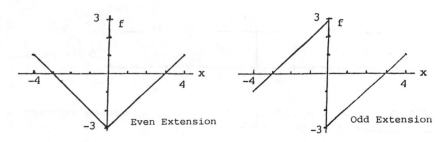

Even Extension   Odd Extension

13. By the hint $f(-x) = g(-x) + h(-x) = g(x) - h(x)$, since g is an even function and f is an odd function. Thus $f(x) + f(-x) = 2g(x)$ and hence $g(x) = (f(x) + f(-x))/2$ defines $g(x)$. $h(x)$ can then be found in a similar manner.

All functions and their derivatives in Problem 14 through 30 are piecxewise continuous on the given intervals and their extensions. Thus the Fourier Theorem applies in all cases.

14. For the cosine series we use the even extension of the function given in Eq.(13) and hence

$$f(x) = \begin{cases} 0 & -2 \leq x < -1 \\ 1+x & -1 \leq x < 0 \end{cases} \quad \text{on the interval } -2 \leq x < 0.$$

However, we don't really need this, as the coefficients in this case are given by Eqs.(7), which just use the original values for $f(x)$ on $0 < x \leq 2$. Applying Eqs.(7) we have $L = 2$ and thus

$$a_0 = (2/2)\int_0^1 (1-x)\,dx + (2/2)\int_1^2 0\,dx = 1/2. \quad \text{Similarly,}$$

$$a_n = (2/2)\int_0^1 (1-x)\cos(n\pi x/2)\,dx = 4[1-\cos(n\pi/2)]/n^2\pi^2 \text{ and}$$

$b_n = 0$. Substituting these values in the Fourier series yields the desired results.
For the sine series, we use Eqs.(8) with $L = 2$. Thus $a_n = 0$ and

$$b_n = (2/2)\int_0^1 (1-x)\sin(n\pi x/2)\,dx = 4[n\pi/2 - \sin(n\pi/2)]/n^2\pi^2.$$

15. The graph of the function to which the series converges
is shown in the figure. Using Eqs.(7) with L = 2 we have
$a_0 = \int_0^1 dx = 1$ and $a_n = \int_0^1 \cos(n\pi x/2)dx = 2\sin(n\pi/2)/n\pi$.
Thus $a_n = 0$ for n even, $a_n = 2/n\pi$ for n = 1,5,9,...and
$a_n = -2/n\pi$ for n = 3,7,11,... . Hence we may write
$a_{2n} = 0$ and $a_{2n-1} = 2(-1)^{n+1}/(2n-1)\pi$, which when
substituted into the series gives the desired answer.

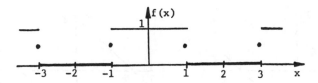

18. The graph of the function to which the series converges
is indicated in the figure.

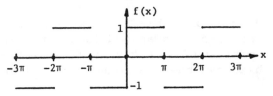

Since we want a sine series, we use Eqs.(8) to find, with
$L = \pi$, that $b_n = (2/\pi)\int_0^\pi \sin nx\,dx = 2[1-(-1)^n]/n\pi$ and thus
$b_n = 0$ for n even and $b_n = 4/n\pi$ for n odd.

20. The graph of the function to which the series converges
is shown in the figure.

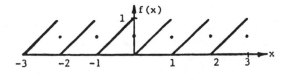

We note that f(x) is specified over its entire
fundamental period (T = 1) and hence we cannot extend f
to make it either an odd or an even function. Using
Eqs.(2) and (3) from Section 10.3 we have (L = 1/2)
$a_0 = 2\int_0^1 x\,dx = 1$, $a_n = 2\int_0^1 x\cos(2n\pi x)dx = 0$ and

$b_n = 2\int_0^1 x\sin(2n\pi x)dx = -1/n\pi$. [Note: We have used the results of Problem 27 of Section 10.2 in writing these integrals. That is, if f(x) is periodic of period T, then every integral of f over an interval of length T has the same value. Thus we integrate from 0 to 1 here, rather than -1/2 to 1/2.] Substituting the above values into Eq.(1) of Section 10.3 yields the desired solution. It can also be observed from the above graph that g(x) = f(x) - 1/2 is an odd function. If Eqs.(8) are used with g(x), then it is found that

$$g(x) = (-1/\pi) \sum_{n=1}^{\infty} \sin(2n\pi x)/n \text{ and thus we obtain the same}$$

series for f(x) as found above.

25a. $b_n = \dfrac{2}{2} \displaystyle\int_0^2 (2-x^2) \sin\dfrac{n\pi x}{2} dx$

$= \dfrac{-2}{n\pi}(2-x^2)\cos\dfrac{n\pi x}{2} - \dfrac{8x}{n^2\pi^2}\sin\dfrac{n\pi x}{2} - \dfrac{16}{n^3\pi^2}\cos\dfrac{n\pi x}{2}\Big|_0^2$

$= \dfrac{4}{n\pi}(1+\cos n\pi) + \dfrac{16}{n^3\pi^3}(1-\cos n\pi)$ and thus

$f(x) = \displaystyle\sum_{n=1}^{\infty} (\dfrac{4n^2\pi^2(1+\cos n\pi) + 16(1-\cos n\pi)}{n^3\pi^3})\sin\dfrac{n\pi x}{2}$

25b.

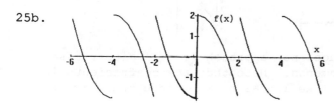

25c.

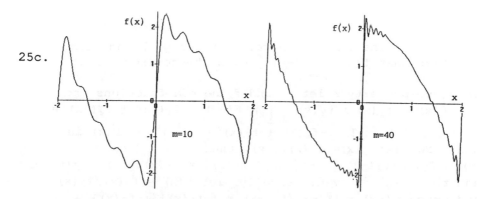

m=10   m=40

28b. For the cosine series (even extension) we have

$a_0 = \dfrac{2}{2}\displaystyle\int_0^1 x dx = \dfrac{1}{2}$

$$a_n = \frac{2}{2}\int_0^1 x\cos\frac{n\pi x}{2}dx = \frac{2x}{n\pi}\sin\frac{n\pi x}{2} + \frac{4}{n^2\pi^2}\cos\frac{n\pi x}{2}\Big|_0^1$$

$$= \frac{2}{n\pi}\sin\frac{n\pi}{2} + \frac{4}{n^2\pi^2}\cos\frac{n\pi}{2} - \frac{4}{n^2\pi^2}, \text{ so}$$

$$g(x) = \frac{1}{4} + \sum_{n=1}^{\infty}\frac{4\cos(n\pi/2) + 2n\pi\sin(n\pi/2) - 4}{n^2\pi^2}\cos\frac{n\pi x}{2}.$$

For the sine series (odd extension) we have

$$b_n = \frac{2}{2}\int_0^1 x\sin\frac{n\pi x}{2}dx = \frac{-2x}{n\pi}\cos\frac{n\pi x}{2} + \frac{4}{n^2\pi^2}\sin\frac{n\pi x}{2}\Big|_0^1$$

$$= \frac{-2}{n\pi}\cos\frac{n\pi}{2} + \frac{4}{n^2\pi^2}\sin\frac{n\pi}{2}, \text{ so}$$

$$h(x) = \sum_{n=1}^{\infty}\frac{4\sin(n\pi/2) - 2n\pi\cos(n\pi/2)}{n\pi^2}\sin\frac{n\pi x}{2}.$$

28c.

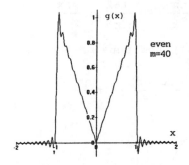 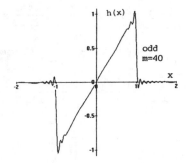

29d. The maximum error does not approach zero in either case, due to Gibb's phenomenon. Note that the coefficients in both series behave like $1/n$ as $n \to \infty$ since there is an n in the numerator.

31. Writing $\int_{-L}^{L}f(x)dx = \int_{-L}^{0}f(x)dx + \int_{0}^{L}f(x)dx$, substituting $x = -y$ into the first integral on the right, and using the fact that $f(x) = -f(-x)$ yields the desired result.

32. To prove property 2 let f_1 and f_2 be odd functions and let $f(x) = f_1(x) \pm f_2(x)$. Then $f(-x) = f_1(-x) \pm f_2(-x) = -f_1(x) \pm [-f_2(x)] = -f_1(x) \mp f_2(x) = -f(x)$, so $f(x)$ is odd. Now let $g(x) = f_1(x)f_2(x)$, then $g(-x) = f_1(-x)f_2(-x) = [-f_1(x)][-f_2(x)] = f_1(x)f_2(x) = g(x)$ and thus $g(x)$ is even. Finally, let $h(x) = f_1(x)/f_2(x)$ and hence $h(-x) = f_1(-x)/f_2(-x) = [-f_1(x)]/[-f_2(x)] = f_1(x)/f_2(x) = h(x)$, which says $h(x)$ is also even. Property 3 is proven in a similar manner.

34. Since $F(x) = \int_0^x f(t)\,dt$ we have

$F(-x) = \int_0^{-x} f(t)\,dt = -\int_0^x f(-s)\,ds$ by letting $t = -s$. If f
is an even function, $f(-s) = f(s)$, we then have
$F(-x) = -\int_0^x f(s)\,ds = -F(x)$ from the original definition of
F. Thus $F(x)$ is an odd function. The argument is
similar if f is odd.

35. Set $x = L/2$ in Eq.(6) of Section 10.3. Since we know f
is continuous at $L/2$, we may conclude, by the Fourier
theorem, that the series will converge to
$f(L/2) = L$ at this point. Thus we have

$L = L/2 + (2L/\pi) \sum_{n=1}^{\infty} (-1)^{n+1}/(2n-1)$, since

$\sin[(2n-1)\pi/2] = (-1)^{n+1}$. Dividing out L and simplifying
yields the desired result.

37a. Multiplying both sides of the equation by $f(x)$ and
integrating from 0 to L gives $\int_0^L [f(x)]^2 dx =$

$\int_0^L [f(x) \sum_{n=1}^{\infty} b_n \sin(n\pi x/L)]\,dx = \sum_{n=1}^{\infty} b_n \int_0^L f(x) \sin(n\pi x/L)\,dx =$

$(L/2) \sum_{n=1}^{\infty} b_n^2$. This result is identical to that of Problem

17 of Section 10.3 if we set $a_n = 0$, $n = 0,1,2,\ldots$. In
a similar manner, it can be shown that

$(2/L)\int_0^L [f(x)]^2 dx = a_0^2/2 + \sum_{n=1}^{\infty} a_n^2$.

37b. Since $f(x) = x$ and $b_n = 2L(-1)^{n+1}/n\pi$, we obtain

$(2/L)\int_0^L [f(x)]^2 dx = (2/L)\int_0^L x^2 dx = 2L^2/3 = \sum_{n=1}^{\infty} b_n^2 =$

$\sum_{n=1}^{\infty} 4L^2/n^2\pi^2 = 4L^2/\pi^2 \sum_{n=1}^{\infty} (1/n^2)$ or $\pi^2/6 = \sum_{n=1}^{\infty} (1/n^2)$.

38. We assume that the extensions of f and f' are piecewise
continuous on $[-2L,2L]$. Since f is an odd periodic
function of fundamental period 4L it follows from
properties 2 and 3 that $f(x)\cos(n\pi x/2L)$ is odd and
$f(x)\sin(n\pi x/2L)$ is even. Thus the Fourier coefficients
of f are given by Eqs.(8) with L replaced by 2L; that is

$a_n = 0$, $n = 0,1,2,\ldots$ and

$b_n = (2/2L)\int_0^{2L} f(x)\sin(n\pi x/2L)\,dx$, $n = 1,2,\ldots$. The

Fourier sine series for f is $f(x) = \sum_{n=1}^{\infty} b_n \sin(n\pi x/2L)$.

39. From Problem 38 we have $b_n = (1/L)\int_0^{2L} f(x)\sin(n\pi x/2L)\,dx$

$= (1/L)\int_0^L f(x)\sin(n\pi x/2L)\,dx + (1/L)\int_L^{2L} f(2L-x)\sin(n\pi x/2L)\,dx$

$= (1/L)\int_0^L f(x)\sin(n\pi x/2L)\,dx - (1/L)\int_L^0 f(s)\sin[n\pi(2L-s)/2L]\,ds$

$= (1/L)\int_0^L f(x)\sin(n\pi x/2L)\,dx - (1/L)\int_0^L f(s)\cos(n\pi)\sin(n\pi s/2L)\,ds$

and thus $b_n = 0$ for n even and

$b_n = (2/L)\int_0^L f(x)\sin(n\pi x/2L)\,dx$ for n odd. The Fourier

series for f is given in Problem 38, where the b_n are

given above.

Section 10.5, Page 587

2. Since the B.C. for this heat conduction problem are
    $u(0,t) = u(100,t) = 0$, $t > 0$, the solution $u(x,t)$ is
    given by Eq.(4) with $\alpha^2 = 1$, cm^2/sec, $L = 40$ cm, and the
    coefficients b_n determined by Eq.(6) with the I.C.
    $u(x,0) = f(x) = x$, $0 \le x \le 20$; $= 40 - x$, $20 \le x \le 40$.

    Thus $b_n = \dfrac{1}{20}[\int_0^{20} x\sin\dfrac{n\pi x}{40}\,dx + \int_{20}^{40}(40-x)\sin\dfrac{n\pi x}{40}\,dx]$

    $= \dfrac{160}{n^2\pi^2}\sin\dfrac{n\pi}{2}$. It follows that

    $u(x,t) = \dfrac{160}{\pi^2}\sum_{n=1}^{\infty}\dfrac{\sin(n\pi/2)}{n^2}e^{-n^2\pi^2 t/1600}\sin\dfrac{n\pi x}{40}$.

7a.

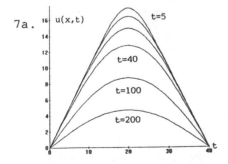

7b.

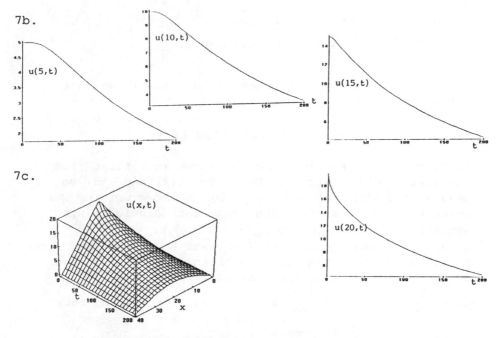

7c.

7d. As in Example 1, the maximum temperature will be at the midpoint and we use just the first term, since the others will be negligible for this temperature. Thus

$$u(20,t) = 1 = \frac{160}{\pi^2} \sin(\pi/2) e^{-\pi^2 t/1600} \sin(20\pi/40).$$ Solving

for t, we obtain $e^{-\pi^2 t/1600} = \pi^2/160$, or

$$t = \frac{1600}{\pi^2} \ln \frac{160}{\pi^2} = 451.60 \text{ sec.}$$

10. Since the B.C. for this heat conduction problem are $u(0,t) = u(20,t) = 0$, t > 0, the solution $u(x,t)$ is given by Eq.(4) with L = 20 cm, and the coefficients b_n determined by Eq.(6) with the I.C. $u(x,0) = f(x) = 100^\circ C$. Thus $b_n = (1/10) \int_0^{20} (100) \sin(n\pi x/20) dx = -200[(-1)^n - 1]/n\pi$ and hence $b_{2n} = 0$ and $b_{2n-1} = 400/(2n-1)\pi$. Substituting these values into Eq.(4) yields

$$u(x,t) = \frac{400}{\pi} \sum_{n=1}^{\infty} \frac{e^{-(2n-1)^2 \pi^2 \alpha^2 t/400}}{2n-1} \sin \frac{(2n-1)\pi x}{20}$$

10b. For aluminum, we have $\alpha^2 = .86$ cm^2/sec (from Table 10.1.1, Sect.10.1) and thus the first two terms give

$$u(10,30) = \frac{400}{\pi} \{e^{-\pi^2(.86)30/400} - \frac{1}{3} e^{-9\pi^2(.86)30/400}\}$$

$$= 67.228^\circ C.$$ If an additional term is used, the

temperature is increased by

$$\frac{80}{\pi}e^{-25\pi^2(.86)30/400} = 3 \times 10^{-6} \text{ degrees C.}$$

11b. Using only one term in the series for $u(x,t)$, we must
solve the equation $5 = (400/\pi)\exp[-\pi^2(.86)t/400]$ for t.
Taking the logarithm of both sides and solving for t
yields $t \cong 400\ln(80/\pi)/\pi^2(.86) = 152.56$ sec.

12a. Since the B.C. are not homogeneous, we must first find
the steady state solution. Using Eqs.(12) and (13) we
have $v'' = 0$ with $v(0) = 0$ and $v(20) = 60$, which has the
solution $v(x) = 3x$. Thus the transient solution $w(x,t)$
satisfies the equations $\alpha^2 w_{xx} = w_t$, $w(0,t) = 0$,
$w(20,t) = 0$ and $w(x,0) = 25 - 3x$, which are obtained from
Eqs.(16) to (18). The solution of this problem is given
by Eq.(4) with the b_n given by Eq.(6):

$$b_n = \frac{1}{10}\int_0^{20}(25-3x)\sin\frac{n\pi x}{20}dx = (70\cos n\pi+50)/n\pi, \text{ and thus}$$

$$u(x,t) = 3x + \sum_{n=1}^{\infty}\frac{70\cos n\pi + 50}{n\pi}e^{-0.86n^2\pi^2t/400}\sin\frac{n\pi x}{20} \text{ since}$$

$\alpha^2 = .86$ for aluminum.

12b.   12c.

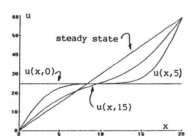

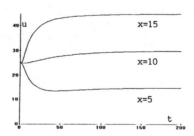

12d. Using just the first term of the sum, we have

$$u(5,t) = 15 - \frac{20}{\pi}e^{-0.86\pi^2t/400}\sin\frac{\pi}{4} = 15 \pm .15. \quad \text{Thus}$$

$$\frac{20}{\pi}e^{-0.86\pi^2t/400}\sin\frac{\pi}{4} = .15, \text{ which yields } t = 160.30 \text{ sec.}$$

To obtain the answer in the text, the first two terms of
the sum must be used, which requires an equation solver
to solve for t. Note that this reduces t by only .01
seconds.

14. Since the B.C. are $u_x(0,t) = u_x(L,t) = 0$, $t > 0$, the solution $u(x,t)$ is given by Eq.(38) with the coefficients c_n determined by Eq.(40). Substituting the I.C. $u(x,0) = f(x) = \sin(\pi x/L)$ into Eq.(40) yields

$$c_0 = (2/L)\int_0^L \sin(\pi x/L)\,dx = 4\pi \text{ and}$$

$$c_n = (2/L)\int_0^L \sin(\pi x/L)\cos(n\pi x/L)\,dx$$

$$= (1/L)\int_0^L \{\sin[(n+1)\pi x/L] - \sin[(n-1)\pi x/L]\}dx$$

$$= (1/\pi)\{[1 - \cos(n+1)\pi]/(n+1) - [1 - \cos(n-1)\pi]/(n-1)\}$$

$$= 0, \text{ n odd}; = -4/(n^2-1)\pi, \text{ n even.} \quad \text{Thus}$$

$$u(x,t) = 2/\pi - (4/\pi)\sum_{n=1}^{\infty} \exp[-4n^2\pi^2\alpha^2 t/L^2]\cos(2n\pi x/L)/(4n^2-1)$$

where we are now summing over even terms by setting $n = 2n$. As $t \to \infty$ we see that all terms in the series decay to zero except the constant term, $2/\pi$. Hence $\lim_{t\to\infty} u(x,t) = 2/\pi$,

16. The steady-state temperature distribution $v(x)$ must satisfy Eq.(12) and also satsify the B.C. $v_x(0) = 0$, $v(L) = 0$. The general solution of $v'' = 0$ is $v(x) = Ax + B$. The B.C. $v_x(0) = 0$ implies $A = 0$ and then $v(L) = 0$ implies $B = 0$, so the steady state solution is $v(x) = 0$.

18a. Substituting $u(x,t) = X(x)T(t)$ into Eq.(1) leads to the two O.D.E. $X'' - \sigma X = 0$ and $T' - \alpha^2 \sigma T = 0$. An argument similar to the one in the text implies that we must have $X(0) = 0$ and $X'(L) = 0$. Also, by assuming σ is real and considering the three cases $\sigma < 0$, $\sigma = 0$, and $\sigma > 0$ we can show that only the case $\sigma < 0$ leads to nontrivial solutions of $X'' - \sigma X = 0$ with $X(0) = 0$ and $X'(L) = 0$. Setting $\sigma = -\lambda^2$, we obtain $X(x) = k_1\sin\lambda x + k_2\cos\lambda x$. Now, $X(0) = 0 \to k_2 = 0$ and thus $X(x) = k_1\sin\lambda x$. Differentiating and setting $x = L$ yields $\lambda k_1\cos\lambda L = 0$. Since $\lambda = 0$ and $k_1 = 0$ lead to $u(x,t) = 0$, we must choose λ so that $\cos\lambda L = 0$, or $\lambda = (2n-1)\pi/2L$, $n = 1,2,3,\ldots$. These values for λ imply that $\sigma = -(2n-1)^2\pi^2/4L^2$ so the solutions $T(t)$ of $T' - \alpha^2\sigma T = 0$ are proportional to $\exp[-(2n-1)^2\pi^2\alpha^2 t/4L^2]$. Combining the above results leads to the desired set of fundamental solutions.

18b. In order to satisfy the I.C. $u(x,0) = f(x)$ we assume that $u(x,t)$ has the form

$$u(x,t) = \sum_{n=1}^{\infty} c_n \exp[-(2n-1)^2 \pi^2 \alpha^2 t/4L^2] \sin[(2n-1)\pi x/2L].$$ The

coefficients c_n are determined by the requirement that

$$u(x,0) = \sum_{n=1}^{\infty} c_n \sin[(2n-1)\pi x/2L] = f(x).$$ Referring to

Problem 39 of Section 10.4 reveals that such a
representation for $f(x)$ is possible if we choose the
coefficients $c_n = (2/L) \int_0^L f(x) \sin[(2n-1)\pi x/2L]dx$.

19. To reduce the problem to one with homogeneous B.C. we
    first find the steady state solution by solving $v'' = 0$,
    subject to the B.C. $v(0) = T$, $v'(L) = 0$. We find that
    $v(x) = T$. As in the text, we write
    $u(x,t) = v(x) + w(x,t)$ and find that w must satisfy the
    equation $\alpha^2 w_{xx} = w_t$, the B.C. $w(0,t) = w_x(L,t) = 0$, and
    the I.C. $w(x,0) = u(x,0) - v(x) = f(x) - T$. Then a
    formal series expansion for $w(x,t)$, and hence $u(x,t)$, can
    be obtained from Problem 18 with $f(x)$ replaced by
    $f(x) - T$ in determining the Fourier coefficients.

21. We must solve $v_1''(x) = 0$, $0 \le x \le a$ and $v_2''(x) = 0$,
    $a \le x \le L$ subject to the B.C. $v_1(0) = 0$, $v_2(L) = T$ and
    the continuity conditions at $x = a$: $v_1(a) = v_2(a)$ and
    $-\kappa_1 A_1 v_1'(a) = -\kappa_2 A_2 v_2'(a)$. The general solutions to the two
    O.D.E. are $v_1(x) = C_1(x) + D_1$ and $v_2(x) = C_2(x) + D_2$. By
    applying the boundary and continuity conditons we may
    solve for C_1, D_1, and C_2 and D_2 to obtain the desired
    solution.

Section 10.6, Page 597

1a. Since the initial velocity is zero, the solution is given
    by Eq.(22) with the coefficients k_n given by Eq.(20).
    Substituting $f(x)$ into Eq.(20) yields

$$k_n = \frac{2}{L}[\int_0^{L/2} \frac{2x}{L} \sin\frac{n\pi x}{L}dx + \int_{L/2}^L \frac{2(L-x)}{L} \sin\frac{n\pi x}{L}dx]$$

$$= \frac{8}{n^2 \pi^2} \sin\frac{n\pi}{2}.$$ Thus Eq. (22) becomes

$$u(x,t) = \frac{8}{\pi^2} \sum_{n=1}^{\infty} \frac{1}{n^2} \sin\frac{n\pi}{2} \sin\frac{n\pi x}{L} \cos\frac{n\pi a t}{L}.$$

1b.

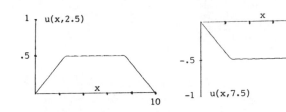 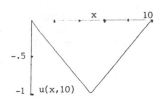

1c.

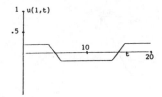 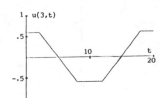 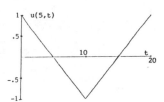

1e. The graphs in part b can best be understood using the
     results of Problem 15. The original triangular shape is
     composed of two similar triangles of 1/2 the height, one
     of which moves to the right and the other to the left.
     Recalling that the series are periodic then gives the
     results shown. The graphs in part c can then be
     visualized from those in part b.

5.   The solution is given by Eq.(18) with the coefficients k_n
     and c_n given by Eqs.(20) and (29). Substituting
     $u(x,0) = f(x) = 0$ into Eq.(20) yields $k_n = 0$,
     $n = 1,2,\ldots$. From Eq.(29) we see that
     $$c_n = (2/n\pi a)\int_0^L g(x)\sin(n\pi x/L)\,dx,\quad n = 1,2,\ldots\ .$$

7a. From Problem 5 we have
     $$c_n = \frac{2}{n\pi a}\left[\int_0^{L/4}\frac{4x}{L}\sin\frac{n\pi x}{L}dx + \int_{L/4}^{3L/4}\sin\frac{n\pi x}{L}dx + \int_{3L}^{L}\frac{4(L-x)}{L}\sin\frac{n\pi x}{L}dx\right]$$

     $$= \frac{8L}{n^3\pi^3 a}\left(\sin\frac{n\pi}{4} + \sin\frac{3n\pi}{4}\right).\quad \text{Substituting this in the}$$
     solution to Problem 5 yields

     $$u(x,t) = \frac{8L}{a\pi^3}\sum_{n=1}^{\infty}\frac{\sin\frac{n\pi}{4} + \sin\frac{3n\pi}{4}}{n^3}\sin\frac{n\pi x}{L}\sin\frac{n\pi a t}{L}.$$

7b.
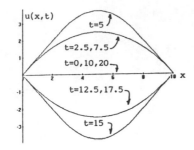

7c.
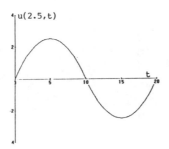

11. Assuming that $u(x,t) = X(x)T(t)$ and substituting for u in
Eq.(1) leads to the pair of O.D.E. $X'' + \sigma X = 0$,
$T'' + a^2\sigma T = 0$. Applying the B.C. $u(0,t) = 0$ and
$u_x(L,t) = 0$ to $u(x,t)$ we see that we must have $X(0) = 0$
and $X'(L) = 0$. By considering the three cases $\sigma < 0$,
$\sigma = 0$, and $\sigma > 0$ it can be shown that nontrivial
solutions of the problem $X'' + \sigma X = 0$, $X(0) = 0$, $X'(L) = 0$
are possible if and only if
$\sigma = (n-1/2)^2\pi^2/L^2 = (2n-1)^2\pi^2/4L^2$, $n = 1,2,\ldots$ and the
corresponding solutions for $X(x)$ are proportional to
$\sin[(2n-1)\pi x/2L]$. Using these values for σ we find that
$T(t)$ is a linear combination of $\sin[(2n-1)\pi at/2L]$ and
$\cos[(2n-1)\pi at/2L]$. Now, the I.C. $u_t(x,0)$ implies that
$T'(0) = 0$ and thus functions of the form
$u_n(x,t) = \sin[(2n-1)\pi x/2L]\cos[(2n-1)\pi at/2L]$, $n = 1,2,\ldots$
satsify the P.D.E. (1), the B.C. $u(0,t) = 0$, $u_x(L,t) = 0$,
and the I.C. $u_t(x,0) = 0$. We now seek a superposition of
these fundamental solutions u_n that also satisfies the
I.C. $u(x,0) = f(x)$. Thus we assume that

$$u(x,t) = \sum_{n=1}^{\infty} c_n\sin[(2n-1)\pi x/2L]\cos[(2n-1)\pi at/2L].$$ The

I.C. now implies that we must have

$$f(x) = \sum_{n=1}^{\infty} c_n\sin[(2n-1)\pi x/2L].$$ From Problem 39 of Section

10.4 we see that $f(x)$ can be represented by such a series
and that

$$c_n = (2/L)\int_0^L f(x)\sin[(2n-1)\pi x/2L]dx, \quad n = 1,2,\dots .$$

Substituting these values into the above series for u(x,t) yields the desired solution.

12a. From Problem 11 we have

$$c_n = \frac{2}{L}\int_{(L-2)/2}^{(L+2)/2}\sin\frac{(2n-1)\pi x}{2L}dx$$

$$= \frac{-4}{(2n-1)\pi}[\cos(\frac{(2n-1)\pi(L+2)}{4L}) - \cos(\frac{(2n-1)\pi(L-2)}{4L})]$$

$$= \frac{8}{(2n-1)\pi}\sin\frac{(2n-1)\pi}{4}\sin\frac{(2n-1)\pi}{2L} \quad \text{using the}$$

trigonometric relations for cos(A ± B). Substituting this value of c_n into u(x,t) in Problem 11 yields the desired solution.

12b.

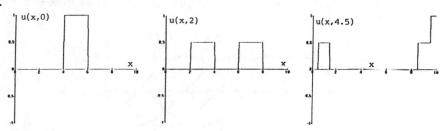

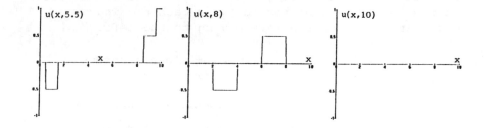

12c.

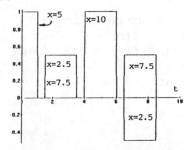

15. Using the chain rule we obtain $u_x = u_\xi \xi_x + u_\eta \eta_x =$
$u_\xi + u_\eta$ since $\xi_x = \eta_x = 1$. Differentiating a second time
gives $u_{xx} = u_{\xi\xi} + 2u_{\xi\eta} + u_{\eta\eta}$. In a similar way we obtain
$u_t = u_\xi \xi_t + u_\eta \eta_t = -au_\xi + au_\eta$, since $\xi_t = -a$, $\eta_t = a$. Thus
$u_{tt} = a^2(u_{\xi\xi} - 2u_{\xi\eta} + u_{\eta\eta})$. Substituting for u_{xx} and u_{tt}
in the wave equation, we obtain $u_{\xi\eta} = 0$. Integrating
both sides of $u_{\xi\eta} = 0$ with respect to η yields
$u_\xi(\xi,\eta) = \gamma(\xi)$ where γ is an arbitrary function of ξ.
Integrating both sides of $u_\xi(\xi,\eta) = \gamma(\xi)$ with respect to ξ
yields $u(\xi,\eta) = \int\gamma(\xi)d\xi + \psi(\eta) = \phi(\xi) + \psi(\eta)$ where $\psi(\eta)$
is an arbitrary function of η and $\int\gamma(\xi)d\xi$ is some
function of ξ denoted by $\phi(\xi)$. Thus
$u(x,t) = u(\xi(x,t),\eta(x,t)) = \phi(x - at) + \psi(x + at)$.

16. The graph of $y = \sin(x-at)$ for the various values of t is
indicated in the figure below. Note that the graph of
$y = \sin x$ is displaced to the right by the distance "at"
for each value of t.

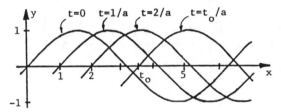

Similarly, the graph of $y = \phi(x + at)$ would be displaced
to the left by a distance "at" for each t. Thus
$\phi(x + at)$ represents a wave moving to the left.

18. Write the equation as $a^2 u_{xx} = u_{tt} + \alpha^2 u$ and assume
$u(x,t) = X(x)T(t)$. This gives $a^2 X''T = XT'' + \alpha^2 XT$,
or $\dfrac{a^2 X''}{X} = \dfrac{T''}{T} + \alpha^2 = \sigma$. The separation constant σ is $-\lambda^2$
using the same arguments as in the text and earlier
problems.

19b. Using the hint and the first equation obtained in part
(a) leads to $\phi(x) + \psi(x) = 2\phi(x) + c = f(x)$ so
$\phi(x) = (1/2)f(x) - c/2$ and $\psi(x) = (1/2)f(x) + c/2$. Hence
$u(x,t) = \phi(x - at) + \psi(x + at) = (1/2)[f(x - at) - c] +$
$(1/2)[f(x + at) + c] = (1/2)[f(x - at) + f(x + at)]$.

19c. Substituting $x + at$ for x in $f(x)$ yields
$$f(x + at) = \begin{cases} 2 & -1 < x + at < 1 \\ 0 & \text{otherwise} \end{cases}$$

Subtracting "at" from both sides of the inequality then yields

$$f(x + at) = \begin{cases} 2 & -1 - at < x < 1 - at \\ 0 & \text{otherwise} \end{cases}$$

20b. From part (a) we have $\psi(x) = -\phi(x)$ and hence
$$\psi(x) = (1/2a)\int_{x_0}^{x} g(\xi)\,d\xi - \phi(x_0).$$

20c. $u(x,t) = \phi(x-at) + \psi(x+at)$
$$= -(1/2a)\int_{x_0}^{x-at} g(\xi)\,d\xi + \phi(x_0) + (1/2a)\int_{x_0}^{x+at} g(\xi)\,d\xi - \phi(x_0)$$

$$= (1/2a)\left[\int_{x_0}^{x+at} g(\xi)\,d\xi - \int_{x_0}^{x-at} g(\xi)\,d\xi\right]$$

$$= (1/2a)\left[\int_{x_0}^{x+at} g(\xi)\,d\xi + \int_{x-at}^{x_0} g(\xi)\,d\xi\right]$$

$$= (1/2a\int_{x-at}^{x+at} g(\xi)\,d\xi.$$

21. Use the result of Problem 10.

24. Substituting $u(r,\theta,t) = R(r)\Theta(\theta)T(t)$ into the P.D.E. yields $R''\Theta T + R'\Theta T/r + R\Theta''T/r^2 = R\Theta T''/a^2$ or equivalently $R''/R + R'/rR + \Theta''/r^2 = T''/a^2T$. In order for this equation to be valid for $0 < r < r_0$, $0 \le \theta \le 2\pi$, $t > 0$, it is necessary that both sides of the equation be equal to the same constant $-\sigma$. Otherwise, by keeping r and θ fixed and varying t, one side would remain unchanged while the other side varied. Thus we arrive at the two equations $T'' + \sigma a^2 T = 0$ and $r^2R''/R + rR'/R + \sigma r^2 = -\Theta''/\Theta$. By an argument similar to the one above we conclude that both sides of the last equation must be equal to the same constant δ. This leads to the two equations $r^2R'' + rR' + (\sigma r^2 - \delta)R = 0$ and $\Theta'' + \delta\Theta = 0$. Since the circular membrane is continuous, we must have $\Theta(2\pi) = \Theta(0)$, which requires $\delta = \mu^2$, μ a non-negative integer. The condition $\Theta(2\pi) = \Theta(0)$ is also known as the periodicity condition. Since we also desire solutions which vary periodically in time, it is clear that the separation constant σ should be positive, $\sigma = \lambda^2$. Thus we arrive at the three equations $r^2R'' + rR' + (\lambda^2 r^2 - \mu^2)R = 0$, $\Theta'' + \mu^2\Theta = 0$, and $T'' + \lambda^2 a^2 T = 0$.

Section 10.7, Page 609

1a. Assuming that $u(x,y) = X(x)Y(y)$ leads to the two O.D.E.
     $X'' - \sigma X = 0$, $Y'' + \sigma Y = 0$. The B.C. $u(0,y) = 0$,
     $u(a,y) = 0$ imply that $X(0) = 0$ and $X(a) = 0$. Thus
     nontrivial solutions to $X'' - \sigma X = 0$ which satisfy these
     boundary conditions are possible only if $\sigma = -(n\pi/a)^2$,
     $n = 1,2\ldots$; the corresponding solutions for $X(x)$ are
     proportional to $\sin(n\pi x/a)$. The B.C. $u(x,0) = 0$ implies
     that $Y(0) = 0$. Solving $Y'' - (n\pi/a)^2 Y = 0$ subject to this
     condition we find that $Y(y)$ must be proportional to
     $\sinh(n\pi y/a)$. The fundamental solutions are then
     $u_n(x,y) = \sin(n\pi x/a)\sinh(n\pi y/a)$, $n = 1,2,\ldots$, which
     satisfy Laplace's equation and the homogeneous B.C. We

     assume that $u(x,y) = \displaystyle\sum_{n=1}^{\infty} c_n \sin(n\pi x/a)\sinh(n\pi y/a)$, where

     the coefficients c_n are determined from the B.C.

     $u(x,b) = g(x) = \displaystyle\sum_{n=1}^{\infty} c_n \sin(n\pi x/a)\sinh(n\pi b/a)$. It follows

     that
     $$c_n \sinh(n\pi b/a) = (2/a)\int_0^a g(x)\sin(n\pi x/a)\,dx, \quad n = 1,2,\ldots .$$

1b. Substituting for $g(x)$ in the equation for c_n we have
     $c_n \sinh(n\pi b/a) = (2/a)\,[\int_0^{a/2} x\sin(n\pi x/a)\,dx +$

     $\int_{a/2}^a (a-x)\sin(n\pi x/a)\,dx] = [4a\,\sin(n\pi/2)]/n^2\pi^2$, $n = 1,2,\ldots,$
     so $c_n = [4a\,\sin(n\pi/2)]/[n^2\pi^2\sinh(n\pi b/a)]$. Substituting
     these values for c_n in the above series yields the
     desired solution.

1c.

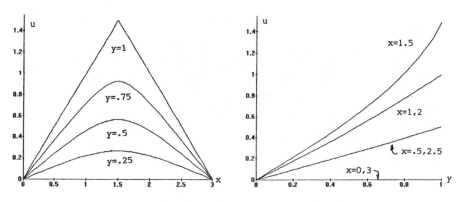

1d.

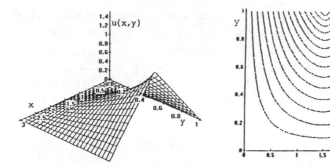

2. In solving the D.E. $Y'' - \lambda^2 Y = 0$, one normally writes
    $Y(y) = c_1 \sinh\lambda y + c_2 \cosh\lambda y$. However, since we have
    $Y(b) = 0$, it is advantageous to rewrite Y as
    $Y(y) = d_1 \sinh\lambda(b-y) + d_2 \cosh\lambda(b-y)$, where d_1, d_2 are also
    arbitrary constants and can be related to c_1, c_2 using
    the appropriate hyperbolic trigonometric identities. The
    important thing, however, is that the second form also
    satisfies the D.E. and thus $Y(y) = d_1 \sinh\lambda(b-y)$ satisfies
    the D.E. and the homogeneous B.C. $Y(b) = 0$. The rest of
    the problem follows the pattern of Problem 1.

3a. Let $u(x,y) = v(x,y) + w(x,y)$, where u, v and w all
    satisfy Laplace's Eq., $v(x,y)$ satisfies the conditions in
    Eq. (4) and $w(x,y)$ satisfies the conditions of Problem 2.

4. Following the pattern of Problem 3, one would consider
    adding the solutions of four problems, each with only one
    non-homogeneous B.C. It is also possible to consider
    adding the solutions of only two problems, each with only
    two non-homogeneous B.C., as long as they involve the
    same variable. For instance, one such problem would be
    $u_{xx} + u_{yy} = 0$, $u(x,0) = 0$, $u(x,b) = 0$, $u(0,y) = k(y)$,
    $u(a,y) = f(y)$, which has the fundamental solutions
    $u_n(x,y) = [c_n \sinh(n\pi x/b) + d_n \cosh(n\pi x/b)]\sin(n\pi y/b)$.

    Assuming $u(x,y) = \sum_{n=1}^{\infty} u_n(x,y)$ and using the B.C.

    $u(0,y) = k(y)$ we obtain $d_n = (2/a)\int_0^a k(y)\sin(n\pi y/b)\,dy$.
    Using the B.C. $u(a,y) = f(y)$ we obtain
    $c_n \sinh(n\pi a/b) + d_n \cosh(n\pi a/b) = (2/a)\int_0^a f(y)\sin(n\pi y/b)\,dy$,
    which can be solved for c_n, since d_n is already known.

5. Using Eq.(19) and following the same arguments as
    presented in the text, we find that $R(r) = k_1 r^n + k_2 r^{-n}$
    and $\Theta(\theta) = c_1 \sin n\theta + c_2 \cos n\theta$, for n a positive integer,

and $u_0(r,\theta) = 1$ for $n = 0$. Since we require that $u(r,\theta)$ be bounded as $r \to \infty$, we conclude that $k_1 = 0$. The fundamental solutions are therefore $u_n(r,\theta) = r^{-n}\cos n\theta$, $v_n(r,\theta) = r^{-n}\sin n\theta$, $n = 1,2,...$ together with $u_0(r,\theta) = 1$. Assuming that u can be expressed as a linear combination of the fundamental solutions we have

$$u(r,\theta) = c_0/2 + \sum_{n=1}^{\infty} r^{-n}(c_n\cos n\theta + k_n\sin n\theta). \quad \text{The B.C.}$$

requires that

$$u(a,\theta) = c_0/2 + \sum_{n=1}^{\infty} a^{-n}(c_n\cos n\theta + k_n\sin n\theta) = f(\theta) \text{ for}$$

$0 \le \theta < 2\pi$. This is precisely the Fourier series representation for $f(\theta)$ of period 2π and thus $a^{-n}c_n = (1/\pi)\int_0^2 f(\theta)\cos n\theta d\theta$, $n = 0,1,2,...$ and

$$a^{-n}k_n = (1/\pi)\int_0^2 f(\theta)\sin n\theta d\theta, \quad n = 1,2... .$$

7. Again we let $u(r,\theta) = R(r)\Theta(\theta)$ and thus we have $r^2 R'' + rR' - \sigma R = 0$ and $\Theta'' + \sigma\Theta = 0$, with $R(0)$ bounded and the B.C. $\Theta(0) = \Theta(\alpha) = 0$. Since $u(r,\theta)$ must be single valued we conclude that $\sigma = \lambda^2$ (λ^2 real) and thus $\Theta(\theta) = c_1\cos\lambda\theta + c_2\sin\lambda\theta$. The B.C. $\Theta(0) = 0 \to c_1 = 0$ and the B.C. $\Theta(\alpha) = 0 \to \lambda = n\pi/\alpha$, $n = 1,2,...$. Substituting these values into Eq.(30) we obtain $R(r) = k_1 r^{n\pi/\alpha} + k_2 r^{-n\pi/\alpha}$. However $k_2 = 0$ since $R(0)$ must be bounded, and thus the fundamental solutions are $u_n(r,\theta) = r^{n\pi/\alpha}\sin(n\pi\theta/\alpha)$. The desired solution may now be formed using previously discussed procedures.

8a. Since neither sinhy nor coshy are bounded as $y \to \infty$, we must write the solution to $Y'' - (n\pi/a)^2 Y = 0$ as $Y(y) = c_1\exp[n\pi y/a] + c_2\exp[-n\pi y/a]$. Thus we must choose $c_1 = 0$ so that $u(x,y) = X(x)Y(y) \to 0$ as $y \to \infty$.

8b. $c_n = \dfrac{2}{a}\int_0^a x(a-x)\sin\dfrac{n\pi x}{a}dx = \dfrac{4a^2}{n^3\pi^3}(1-\cos n\pi)$

8c. Using just the first term and letting $a = 5$, we have $u(x,y) = \dfrac{200}{\pi^3}e^{-\pi y/5}\sin\dfrac{\pi x}{5}$, which, for a fixed y, has a maximum at $x = 5/2$ and thus we need to find y such that

$u(5/2,y) = \dfrac{200}{\pi^3}e^{-\pi y/5} = .1.$ Taking the logarithm of both

sides and solving for y yields $y_0 = 6.6315$. With an equation solver, more terms can be used. However, to four decimal places, three terms yield the same result as above.

13a. Assuming that $u(x,y) = X(x)Y(y)$ and substituting into
   Eq.(1) leads to the two O.D.E. $X'' - \sigma X = 0$, $Y'' + \sigma Y = 0$.
   The B.C. $u(x,0) = 0$, $u_y(x,b) = 0$ imply that $Y(0) = 0$ and
   $Y'(b) = 0$. For nontrivial solutions to exist for
   $Y'' + \sigma Y = 0$ with these B.C. we find that σ must take the
   values $(2n-1)^2\pi^2/b^2$, $n = 1,2,\ldots$; the corresponding
   solutions for $Y(y)$ are proportional to $\sin[(2n-1)\pi y/b]$.
   Solutions to $X'' - [(2n-1)^2\pi^2/b^2]X = 0$ are of the form
   $X(x) = A\sinh[(2n-1)\pi x/2b] + B\cosh[(2n-1)\pi x/2b]$. However,
   the boundary condition $u(0,y) = 0$ implies that
   $X(0) = B = 0$. It follows that the fundamental solutions
   are $u_n(x,y) = c_n\sinh[(2n-1)\pi x/2b]\sin[(2n-1)\pi y/2b]$,
   $n = 1,2,\ldots$. To satisfy the remaining B.C. at $x = a$ we
   assume that we can represent the solution $u(x,y)$ in the

   form $u(x,y) = \displaystyle\sum_{n=1}^{\infty} c_n\sinh[(2n-1)\pi x/2b]\sin[(2n-1)\pi y/2b]$.

   The coefficients c_n are determined by the B.C.

   $u(a,y) = \displaystyle\sum_{n=1}^{\infty} c_n\sinh[(2n-1)\pi a/2b]\sin[(2n-1)\pi y/2b] = f(y)$.

   By properly extending f as a periodic function of period
   4b as in Problem 39, Section 10.4, we find that the
   coefficients c_n are given by
   $c_n\sinh[(2n-1)\pi a/2b] = (2/b)\displaystyle\int_0^b f(y)\sin[(2n-1)\pi y/2b]\,dy$,

   $n = 1,2,\ldots$.

CHAPTER 11

Section 11.1, Page 622

2. Since the B.C. at $x = 1$ is nonhomogeneous, the B.V.P. is nonhomogeneous.

4. The D.E. may be written $y'' + (\lambda - x^2)y = 0$ and is thus homogeneous, as are both B.C.

5. Since the D.E. contains the nonhomogeneous term 1, the B.V.P. is nonhomogeneous.

7b. Eq.(iii) is linear and separable. Using the latter approach we have $d\mu/\mu = [(Q-P')/P]dx$ and thus $\ln\mu = \int_{x_0}^{x} [Q(s)/P(s)]ds - \ln P$. Taking the exponential of both sides yields Eq.(iv). The choice of x_0 simply alters the constant of integration, which is immaterial here.

9. Since $P(x) = x^2$ and $Q(x) = x$, we find that $\mu(x) = (1/x^2)\exp[\int_{x_0}^{x} (s/s^2)ds] = k/x$, where k is an arbitrary constant which may be set equal to 1. It follows that Bessel's equation takes the form $(xy')' + (x-\upsilon^2/x)y = 0$.

Section 11.2, Page 629

1. If $\lambda < 0$, the general solution of the D.E. is $y = c_1\sinh\sqrt{\mu}\,x + c_2\cosh\sqrt{\mu}\,x$ where $-\lambda = \mu$. The two B.C. require that $c_2 = 0$ and $c_1 = 0$ so $\lambda < 0$ is not an eigenvalue. If $\lambda = 0$, the general solution of the D.E. is $y = c_1 + c_2 x$. The B.C. require that $c_1 = 0$, $c_2 = 0$ so $\lambda = 0$ is not an eigenvalue.
    If $\lambda > 0$, the general solution of the D.E. is $y = c_1\sin\sqrt{\lambda}\,x + c_2\cos\sqrt{\lambda}\,x$. The B.C. require that $c_2 = 0$, $\sqrt{\lambda}\,c_1\cos\sqrt{\lambda} = 0$. The second condition is satisfied for $\lambda \neq 0$ and $c_1 \neq 0$ if $\lambda = [(2n-1)\pi/2]^2$, $n = 1,2,\dots$. Thus the eigenvalues are $\lambda_n = [(2n-1)\pi/2]^2$, $n = 1,2,\dots$ with corresponding eigenfunctions $\phi_n(x) = \sin[(2n-1)\pi x/2]$, $n = 1,2,\dots$.

4. Note that the problem is identical to Problem 1 with λ replaced by $-\lambda$.

5. If $\lambda = 0$, the general solution of the D.E. is $y = c_1 + c_2 x$. The B.C. $y(0) + y'(0) = 0$ requires $c_1 + c_2 = 0$ and the B.C. $y(1) = 0$ requires $c_1 + c_2 = 0$ and thus $\lambda = 0$ is an eigenvalue with corresponding eigenfunction $\phi_0(x) = 1-x$.

    If $\lambda < 0$, set $-\lambda = \mu^2$ to obtain $y = c_1 \cos\mu x + c_2 \sin\mu x$. In this case the B.C. require $c_1 + \mu c_2 = 0$ and $c_1 \cos\mu + c_2 \sin\mu = 0$ which yields nontrivial solutions for c_1 and c_2 (i.e., $c_1 = -\mu c_2$) if and only if $\tan\mu = \mu$. By plotting on the same graph $f(\mu) = \mu$ and $g(\mu) = \tan\mu$, we see that they intersect at $\mu_0 = 0$ ($\mu = 0 \rightarrow \lambda = 0$, which has already been discussed), $\mu_1 \cong 4.4934$ (which is just to the left of the vertical asymptote of $\tan\mu$ at $\mu = 3\pi/2$, $\mu_2 \cong 7.72525$ (which is just to the left of the vertical asymptote of $\tan\mu$ at $\mu = 5\pi/2$) and for larger values $\mu_n \cong (2n+1)\pi/2$. Since $\lambda_n = -\mu_n^2$, we have $\lambda_1 \cong -20.1906$, $\lambda_2 = -59.6795$, $\lambda_n \cong -(2n+1)^2\pi^2/4$ and $\phi_n = \sin\mu_n x - \mu_n \cos\mu_n x$.

    If $\lambda > 0$, the general solution of the D.E. is $y(x) = c_1 \cosh\sqrt{\lambda} x + c_2 \sinh\sqrt{\lambda} x$. The B.C. respectively require that $c_1 + \sqrt{\lambda} c_2 = 0$ and $c_1 \cosh\sqrt{\lambda} + c_2 \sinh\sqrt{\lambda} = 0$ and thus λ must satisfy $\tanh\sqrt{\lambda} = \sqrt{\lambda}$ in order to have nontrivial solutions. The only solution of this equation is $\lambda = 0$ and thus there are no positive eigenvalues.

8. If $\lambda > 0$, the general solution of the D.E. is $y = c_1 \sin\sqrt{\lambda} x + c_2 \cos\sqrt{\lambda} x$. The B.C. require that $c_2 - \sqrt{\lambda} c_1 = 0$, $(\sin\sqrt{\lambda} + \sqrt{\lambda} \cos\sqrt{\lambda})c_1 + (\cos\sqrt{\lambda} - \sqrt{\lambda} \sin\sqrt{\lambda})c_2 = 0$. In order to have nontrivial solutions λ must satisfy $(\lambda-1)\sin\sqrt{\lambda} - 2\sqrt{\lambda} \cos\sqrt{\lambda} = 0$. In this case $c_2 = \sqrt{\lambda} c_1$ and thus $\phi_n = \sin\sqrt{\lambda_n} x + \sqrt{\lambda_n} \cos\sqrt{\lambda_n} x$. If $\lambda \neq 1$, the eigenvalue equation is equivalent to $\tan\sqrt{\lambda} = 2\sqrt{\lambda}/(\lambda-1)$ and thus by graphing $f(\sqrt{\lambda}) = \tan\sqrt{\lambda}$ and $g(\sqrt{\lambda}) = 2\sqrt{\lambda}/(\lambda-1)$ we can estimate the eigenvalues. Since $g(\sqrt{\lambda})$ has a vertical asymptote at $\lambda = 1$ and $f(\sqrt{\lambda})$ has a vertical asymptote at $\sqrt{\lambda} = \pi/2$, we see that $1 < \sqrt{\lambda_1} < \pi/2$.

By interating numerically, we find $\sqrt{\lambda_1} \cong 1.30655$ and thus $\lambda_1 \cong 1.7071$. The second eigenvalue will lie to the right of π, the second zero of $\tan\sqrt{\lambda}$. Again iterating numerically, we find $\sqrt{\lambda_2} \cong 3.6732$ and thus $\lambda_2 \cong 13.4924$. For large values of n, we see from the graph that $\sqrt{\lambda_n} \cong (n-1)\pi$, which are the zeros of $\tan\sqrt{\lambda}$. Thus $\lambda_n \cong (n-1)^2\pi^2$ for large n. For $\lambda \le 0$, the discussion follows the pattern of earlier problems.

10a. Assuming $y = s(x)u$, we have $y' = s'u + su'$ and $y'' = s''u + 2s'u' + su''$ and thus the D.E. becomes $su'' + (2s'+4s)u' + [s'' + 4s' + (4+9\lambda)s]u = 0$. Setting $2s' + 4s = 0$ we find $s(x) = e^{-2x}$ and the D.E. becomes $u'' + 9\lambda u = 0$. The B.C. $y(0) = 0$ yields $s(0)u(0) = 0$, or $u(0) = 0$ since $s(0) \ne 0$. The B.C. at L is $y'(L) = s'(L)u(L) + s(L)u'(L) = e^{-2L}(-2u(L) + u'(L)) = 0$ and thus $u'(L) - 2u(L) = 0$. Thus the B.V.P. satisfied by $u(x)$ is $u'' + 9\lambda u = 0$, $u(0) = 0$, $u'(1) - 2u(L) = 0$.

If $\lambda < 0$, the general solution of the D.E. $u'' + 9\lambda u = 0$ is $u = c_1\sinh 3\mu x + c_2\cosh 3\mu x$ where $-\lambda = \mu^2$. The B.C. require that $c_2 = 0$, $c_1(3\mu\cosh 3\mu L - 2\sinh 3\mu L) = 0$. In order to have nontrivial solutions μ must satisfy the equation $3\mu/2 = \tanh 3\mu L$. A graphical analysis reveals that for $L \le 1/2$ this equation has no solutions for $\mu \ne 0$ so there are no negative eigenvalues for $L \le 1/2$. If $L > 1/2$ there is one solution and hence one negative eigenvalue with eigenfuction $\phi_{-1}(x) = e^{-2x}\sinh 3\mu x$.

If $\lambda = 0$, the general solution of the D.E. $u'' + 9\lambda u = 0$ is $u = c_1 + c_2 x$. The B.C. require that $c_1 = 0$, $c_2(1-2L) = 0$ so nontrivial solutions are possible only if $L = 1/2$. In this case the eigenfuction is $\phi_0(x) = xe^{-2x}$.

If $\lambda > 0$, the general solution of the D.E. $u'' + 9\lambda u = 0$ is $u = c_1\sin 3\sqrt{\lambda} x + c_2\cos\sqrt{\lambda} x$. The B.C. require that $c_2 = 0$, $c_1(3\sqrt{\lambda}\cos 3\sqrt{\lambda} L - 2\sin 3\sqrt{\lambda} L) = 0$. In order to have nontrivial solutions λ must satisfy the equation $\sqrt{\lambda} = (2/3)\tan 3\sqrt{\lambda} L$. A graphical analysis reveals that there is an infinite number of solutions to this eigenvalue equation. Thus the eigenfunctions are $\phi_n(x) = e^{-2x}\sin 3\sqrt{\lambda_n} x$ where the eigenvalues λ_n satisfy $\sqrt{\lambda_n} = (2/3)\tan 3\sqrt{\lambda_n} L$.

12. This is an Euler equation. If $\lambda = 1$ the general solution
of the D.E. is $y = c_1 x + c_2 x \ln x$ and the B.C. require that
$c_1 = c_2 = 0$ and thus $\lambda = 1$ is not an eigenvalue. If
$\lambda \neq 1$, $y = c_1 x + c_2 x^\lambda$ is the general solution and the
B.C. require that $c_1 + c_2 = 0$ and
$2c_1 + c_2 2^\lambda - (c_1 + \lambda c_2 2^{\lambda-1}) = 0$. Thus nontrivial solutions
exist if and only if $\lambda = 2(1-2^{-\lambda})$. The graphs of
$f(\lambda) = \lambda$ and $g(\lambda) = 2(1-2^{-\lambda})$ intersect only at $\lambda = 1$
(which has already been discussed) and $\lambda = 0$. Thus the
only eigenvalue is $\lambda = 0$ with corresponding eigenfunction
$\phi(x) = x - 1$ (since $c_1 = -c_2$).

14a. For positive λ, the general solution of the D.E. is
$y = c_1 \sin\sqrt{\lambda}\, x + c_2 \cos\sqrt{\lambda}\, x$. The B.C. require that
$\sqrt{\lambda}\, c_1 + \alpha c_2 = 0$, $c_1 \sin\sqrt{\lambda} + c_2 \cos\sqrt{\lambda} = 0$. Nontrivial
solutions exist if and only if $\sqrt{\lambda} \cos\sqrt{\lambda} - \alpha \sin\sqrt{\lambda} = 0$.
If $\alpha = 0$ this equation is satisfied by the sequence
$\lambda_n = [(2n-1)\pi/2]^2$, $n = 1,2,\ldots$. If $\alpha \neq 0$, λ must
satisfy the equation $\sqrt{\lambda}/\alpha = \tan\sqrt{\lambda}$. A plot of the
graphs of $f(\sqrt{\lambda}) = \sqrt{\lambda}/\alpha$ and $g(\sqrt{\lambda}) = \tan\sqrt{\lambda}$ reveals
that there is an infinite sequence of postive eigenvalues
for $\alpha < 0$ and $\alpha > 0$.

14b. By procedures shown previously, the cases $\lambda < 0$ and
$\lambda = 0$, when $\alpha < 1$, lead to only the trivial solution and
thus by part a all real eigenvalues are positive. For
$0 < \alpha < 1$, the graphs of $f(\sqrt{\lambda})$ and $g(\sqrt{\lambda})$ (see part a)
intersect once on $0 < \sqrt{\lambda} < \pi/2$. As α approaches 1 from
below, the slope of $f(\sqrt{\lambda})$ decreases and thus the
intersection point approaches zero.

15. Using the D.E. for ϕ_m and following the hint yields:
$\int_0^L \phi_m'' \phi_n dx + \lambda_m \int_0^L \phi_m \phi_n dx = 0$. Integrating the first term by
parts yields: $\phi_m' \phi_n \Big|_0^L - \int_0^L \phi_m' \phi_n' dx = -\lambda_m \int_0^L \phi_m \phi_n dx$. Upon
utilizing the B.C. the first term on the left vanishes
and thus $\int_0^L \phi_n' \phi_m' dx = \lambda_m \int_0^L \phi_m \phi_n dx$. Similarly, the D.E. for
ϕ_n yields $\int_0^L \phi_m' \phi_n' dx = \lambda_n \int_0^L \phi_n \phi_m dx$ and thus
$(\lambda_n - \lambda_m) \int_0^L \phi_m \phi_n dx = 0$. If $\lambda_n \neq \lambda_m$ the desired result
follows.

16b. The general solution of the D.E. is
     $y = c_1\sin\mu x + c_2\cos\mu x + c_3\sinh\mu x + c_4\cosh\mu x$ where $\lambda = \mu^4$.
     The B.C. require that $c_2 + c_4 = 0$, $-c_2 + c_4 = 0$,
     $c_1\sin\mu L + c_2\cos\mu L + c_3\sinh\mu L + c_4\cosh\mu L = 0$, and
     $c_1\cos\mu L - c_2\sin\mu L + c_3\cosh\mu L + c_4\sinh\mu L = 0$. The first
     two equations yield $c_2 = c_4 = 0$, and the last two have
     nontrivial solutions if and only if
     $\sin\mu L\cosh\mu L - \cos\mu L\sinh\mu L = 0$. In this case the third
     equation yields $c_3 = -c_1\sin\mu L/\sinh\mu L$ and thus the desired
     eigenfunctions are obtained. The quantity μL can be
     approximated by finding the intersection of $f(x) = \tan x$
     and $g(x) = \tanh x$, where $x = \mu L$. The first intersection
     is at $x \cong 3.9266$, which gives $\lambda_1 \cong 237.72/L^4$ and the
     second intersection is at $x \cong 7.0686$, which gives $\lambda_2 \cong$
     $2.496.5/L^4$.

Section 11.3, Page 641

1.   In Problem 1 of Section 11.2 we found the eigenfuctions
     to be $\sin[(2n-1)\pi x/2]$, $n = 1,2,\ldots$ and thus, by Eq.(20),
     we must choose k_n so that $\int_0^1 \{k_n\sin[(2n-1)\pi x/2]\}^2 dx = 1$,
     since the weight function $r(x) = 1$ (by comparing the D.E.
     to Eq.(1)). Evaluating the integral yields $k_n^2/2 = 1$ and
     thus $k_n = \sqrt{2}$ and the desired normalized eigenfuctions
     are obtained.

3.   Note here that $\phi_0(x) = 1$ satisifes Eq.(20) and hence it
     is already normalized.

5.   From Problem 9 of Section 11.2 we have $e^x\sin n\pi x$,
     $n = 1,2,\ldots$ as the eigenfunctions and thus k_n must be chosen
     so that $\int_0^1 r(x)k_n^2 e^{2x}\sin^2 n\pi x\, dx = 1$. To determine $r(x)$, we must
     write the D.E. in the form of Eq.(1). That is, we multiply
     the D.E. by $r(x)$ to obtain
     $ry'' - 2ry' + ry + \lambda ry = 0$. Now choose r so that
     $(ry')' = ry'' - 2ry'$, which yields $r' - 2r = 0$ or $r(x) = e^{-2x}$.
     Thus the above integral becomes $\int_0^1 k_n^2\sin^2 n\pi x\, dx = 1$ and
     $k_n = \sqrt{2}$. Hence $\phi_n(x) = \sqrt{2}\,e^x\sin n\pi x$ are the normalized
     eigenfunctions.

7. Using Eq.(34) with $r(x) = 1$, we find that the
    coefficients of the series (32) are determined by
    $$a_n = (f, \phi_n) = \sqrt{2} \int_0^1 x \sin[(2n-1)\pi x/2] dx$$
    $$= (4\sqrt{2}/(2n-1)^2 \pi^2) \sin(2n-1)\pi/2. \quad \text{Thus Eq.(32) yields}$$
    $$f(x) = \frac{4\sqrt{2}}{\pi^2} \sum_{n=1}^{\infty} \frac{(-1)^{n-1}}{(2n-1)^2} \sqrt{2} \sin[(2n-1)\pi x/2], \quad 0 \le x \le 1,$$
    which agrees with the expansion using the approach
    developed in Problem 39 of Section 10.4.

10. In this case $\phi_n(x) = (\sqrt{2}/\alpha_n) \cos\sqrt{\lambda_n} x$, where
    $\alpha_n = (1 + \sin^2\sqrt{\lambda_n})^{1/2}$. Thus Eq.(34) yields
    $$a_n = (\sqrt{2}/\alpha_n) \int_0^1 \cos\sqrt{\lambda_n} x dx = \sqrt{2} \sin\sqrt{\lambda_n}/\alpha_n\sqrt{\lambda_n}.$$

14. In this case $L[y] = y'' + y' + 2y$ is not of the form shown
    in Eq.(3) and thus the B.V.P. is not self adjoint.

17. In this case $L[y] = [(1+x^2)y']' + y$ and thus the D.E. has
    the form shown in Eq.(3). However, the B.C. are not
    separated and thus we must determine by integration
    whether Eq.(8) is satisfied. Therefore, for u and v
    satisfying the B.C., integration by parts yields the
    following:
    $$(L[u], v) = \int_0^1 \{[(1+x^2)u']' + u\} v dx = vu'(1+x^2)\Big|_0^1 - \int_0^1 \{(1+x^2)v'u' + uv\} dx$$
    $$= vu'(1+x^2)\Big|_0^1 - uv'(1+x^2)\Big|_0^1 + \int_0^1 \{[(1+x^2)v']' + v\} u dx$$
    $$= (u, L[v])$$
    since the integrated terms add to zero with the given
    B.C. Thus the B.V.P. is self-adjoint.

21a. Substituting ϕ for y in the D.E., multiplying both sides
     by ϕ, and integrating form 0 to 1 yields
     $$\lambda \int_0^1 r\phi^2 dx = \int_0^1 \{-[p(x)\phi']'\phi + q(x)\phi^2\} dx. \quad \text{Integrating the}$$
     first term on the right side once by parts, we obtain
     $$\lambda \int_0^1 r\phi^2 dx = -p(1)\phi'(1)\phi(1) + p(0)\phi'(0)\phi(0) + \int_0^1 (p\phi'^2 + q\phi^2) dx.$$
     If $a_2 \neq 0$, $b_2 \neq 0$, then $\phi'(1) = -b_1\phi(1)/b_2$ and
     $\phi'(0) = -a_1\phi(0)/a_2$ and the result follows. If $a_2 = 0$,
     then $\phi(0) = 0$ and the boundary term at 0 will be missing.
     A similar result is obtained if $b_2 = 0$.

21b. Note that $p(x)$ is required to be strictly positive for $0 \leq x \leq 1$.

23a. Using $\phi(x) = u(x) + iv(x)$ in Eq.(4) we have $L[\phi] = L[u(x) + iv(x)] = \lambda r(x)[u(x)+iv(x)]$. Using the linearity of L and the fact that λ and $r(x)$ are real we have $L[u(x)] + iL[v(x)] = \lambda r(x)u(x) + i\lambda r(x)v(x)$. Equating the real and imaginary parts shows that both u and v satisfy Eq.(1). The B.C. Eq.(2) are also satisfied by both u and v, using the same arguments, and thus both u and v are eigenfunctions.

23c. From part b we have $v(x) = cu(x)$ and thus $\phi(x) = u(x) + icu(x) = (1+ic)u(x)$.

24. If $\lambda = 1$, the general solution to the D.E. is $y = c_1 x + c_2 x \ln x$. The B.C. require that $c_1 = 0$, $2c_1 + 2(\ln 2)c_2 = 0$ so $c_1 = c_2 = 0$ and $\lambda = 1$ is not an eigenvalue. If $\lambda \neq 1$, the general solution to the D.E. is $y = c_1 x + c_2 x^\lambda$. The B.C. require that $c_1 + c_2 = 0$, $2c_1 + 2^\lambda c_2 = 0$. Nontrivial solutions exist if and only if $2^\lambda - 2 = 0$. If λ is real, this equation has no solution (other than $\lambda = 1$) and again $y = 0$ is the only solution to the boundary value problem. Suppose that $\lambda = a + bi$ with $b \neq 0$. Then $2^\lambda = 2^{a+bi} = 2^a 2^{bi} = 2^a \exp(ib\ln 2)$, which upon substitution into $2^\lambda = 2$ yields the equation $\exp(ib\ln 2) = 2^{1-a}$. Since $e^{ix} = \cos x + i\sin x$, it follows that $a = 1$ and $b(\ln 2) = 2n\pi$ or $b = 2n\pi/\ln 2$, $n = \pm 1, \pm 2, \ldots$. Thus the only eigenvalues of the problem are $\lambda_n = 1 + i(2n\pi/\ln 2)$, $n = \pm 1, \pm 2, \ldots$.

25b. For $\lambda \leq 0$, there are no eigenfuctions. For $\lambda > 0$ the general solution of the D.E. is $y = c_1 + c_2 x + c_3 \sin\sqrt{\lambda}\, x + c_4 \cos\sqrt{\lambda}\, x$. The B.C. require that $c_1 + c_4 = 0$, $c_4 = 0$, $c_1 + c_2 L + c_3 \sin\sqrt{\lambda}\, L + c_4 \cos\sqrt{\lambda}\, L = 0$, and $c_2 + \sqrt{\lambda}\, c_3 \cos\sqrt{\lambda}\, L - \sqrt{\lambda}\, c_4 \sin\sqrt{\lambda}\, L = 0$. Thus $c_1 = c_4 = 0$ and for nontrivial solutions to exist λ must satisfy the equation $\sqrt{\lambda}\, L\cos\sqrt{\lambda}\, L - \sin\sqrt{\lambda}\, L = 0$. In this case $c_2 = (-\sqrt{\lambda}\cos\sqrt{\lambda}\, L)(c_3)$ and it follows that the eigenfunction ϕ_1 is given by

$\phi_1(x) = \sin(\sqrt{\lambda_1}\, x) - \sqrt{\lambda_1}\, x\cos\sqrt{\lambda_1}\, L$ where λ_1 is the smallest positive solution of the equation $\sqrt{\lambda}\, L = \tan\sqrt{\lambda}\, L$. A graphical or numerical estimate of λ_1 reveals that $\lambda_1 \cong (4.4934)^2/L^2$.

25c. The eigenvalue equation is $2(1-\cos x) = x\sin x$, where $x = \sqrt{\lambda}\, L$. Graphing $f(x) = 2(1-\cos x)$ and $g(x) = x\sin x$ we see that there is an intersection for $6 < x < 7$. Since both $f(x)$ and $g(x)$ are zero for $x = 2\pi$, this then is the precise root and thus $\lambda_1 = (2\pi)^2/L^2$. In addition, it appears there might be an intersection for $0 < x < 1$. Using a Taylor series representation for $f(x)$ and $g(x)$ about $x = 0$, however, shows there is no intersection for $0 < x < 1$. Of course $x = 0$ is also an intersection, which yields $\lambda = 0$, which gives the trivial solution and hence $\lambda = 0$ is not an eigenvalue.

Section 11.4, Page 653

1. We must first find the eigenvalues and normalized eigenfunctions of the associated homogeneous problem $y'' + \lambda y = 0$, $y(0) = 0$, $y(1) = 0$. This problem has the solutions $\phi_n(x) = k_n\sin n\pi x$, for $\lambda_n = n^2\pi^2$, $n = 1, 2, \ldots$. Choosing k_n so that $\int_0^1 \phi_n^2 dx = 1$ we find $k_n = \sqrt{2}$. Hence the solution of the original nonhomogeneous problem is given by $y = \sum_{n=1}^{\infty} b_n\phi_n(x)$, where the coefficients b_n are found from Eq.(12), $b_n = c_n/(\lambda_n - 2)$ where c_n is given by $c_n = \sqrt{2}\int_0^1 x\sin n\pi x\, dx$ (Eq.9). [Note that the original problem can be written as $-y'' = 2y + x$ and therefore comparison with Eq.(1) yields $r(x) = 1$ and $f(x) = x$]. Integrating the expression for c_n by parts yields $c_n = \sqrt{2}\,(-1)^{n+1}/n\pi$ and thus $y = \sum_{n=1}^{\infty} \dfrac{\sqrt{2}\,(-1)^{n+1}}{(n^2\pi^2 - 2)\, n\pi}\, \sqrt{2}\,\sin n\pi x$.

2. From Problem 1 of Section 11.3 we have $\phi_n = \sqrt{2}\,\sin[(2n-1)\pi x/2]$ for $\lambda_n = (2n-1)^2\pi^2/4$ and from Problem 7 of that section we have $c_n = 4\sqrt{2}\,(-1)^{n-1}/(2n-1)^2\pi^2$. Substituting these values

into $b_n = c_n/(\lambda_n - 2)$ and $y = \sum\limits_{n=1}^{\infty} b_n \phi_n$ yields the desired result.

3. Referring to Problem 3 of Section 11.3 we have

$y = b_0 + \sum\limits_{n=1}^{\infty} b_n(\sqrt{2}\cos n\pi x)$, where $b_n = c_n/(\lambda_n - 2)$ for

$n = 0,1,2,\ldots$. The rest of the calculations follow those of Problem 1.

5. Note that the associated eigenvalue problem is the same as for Problem 1 and that $|1-2x| = 1-2x$ for $0 \le x \le 1/2$ while $|1-2x| = 2x-1$ for $1/2 \le x \le 1$.

10. Since $\mu = \pi^2$ is an eigenvalue of the corresponding homogeneous equation, Theorem 11.4.1 tells us that a solution will exist only if $-(a+x)$ is orthogonal to the corresponding eigenfunction $\sqrt{2}\sin\pi x$. Thus we require $\int_0^1 (a+x)\sin\pi x\,dx = 0$, which yields $a = -1/2$. With $a = -1/2$, we find that $Y = (x-1/2)/\pi^2$ and $y_c = c\sin\pi x + d\cos\pi x$ by methods of Chapter 3. Setting $y = y_c + Y$ and choosing d to satisfy the B.C. we obtain the desired family of solutions.

11. Note that in this case $\mu = 4\pi^2$ and $\phi_2 = \sqrt{2}\sin 2\pi x$ are the eigenvalue and eigenfunction respectively of the corresponding homogeneous equation.

12. In this case a solution will exist only if $-a$ is orthogonal to $\sqrt{2}\cos\pi x$., that is if $\int_0^1 a\cos\pi x\,dx = 0$. Since this condition is valid for all a, a family of solutions exists.

14. Since $\sum\limits_{n=1}^{\infty} c_n\phi_n(x)$ converges to zero we have $\sum\limits_{n=1}^{\infty} c_n\phi_n(x) = 0$. Multiplying and integrating as suggested yields

$\int_0^1 [\sum\limits_{n=1}^{\infty} c_n\phi_n(x)] r(x)\phi_m(x)\,dx = 0$ or

$\sum\limits_{n=1}^{\infty} c_n\int_0^1 r(x)\phi_n(x)\phi_m(x)\,dx = 0$. The integral that multiplies

c_n is just δ_{nm} [Eq. (22) of Section 11.3]. Thus the infinite sum becomes c_m and the last equation yields $c_m = 0$.

18. A twice differentiable function v satisfying the boundary conditions can be found by assuming that $v = ax + b$. Thus $v(0) = b = 1$ and $v(1) = a + 1$ while $v'(1) = a$. Hence $2a + 1 = -2$ or $a = -3/2$ and $v(x) = 1 - 3x/2$. Assuming $y = u + v$ we have $(u+v)'' + 2(u+v) = u'' + 2u + 2(1-3x/2) = 2 - 4x$ or $u'' + 2u = -x$, $u(0) = 0$, $u(1) + u'(1) = 0$ which is the same as Example 1 of the text.

19. From Eq.(30) we assume $u(x,t) = \sum_{n=1}^{\infty} b_n(t)\phi_n(x)$, where the ϕ_n are the eigenfunctions of the related eigenvalue problem $y'' + \lambda y = 0$, $y(0) = 0$, $y'(1) = 0$ and the $b_n(t)$ are given by Eq.(42). From Problem 1, Section 11.3 we have $\phi_n = \sqrt{2}\sin[(2n-1)\pi x/2]$ and $\lambda_n = (2n-1)^2\pi^2/4$. To evaluate Eq.(42) we need to calculate
$$\alpha_n = \int_0^1 \sin(\pi x/2)\sqrt{2}\sin[(2n-1)\pi x/2 dx \quad [\text{Eq.(41) with}$$
$r(x) = 1$ and $f(x) = \sin(\pi x/2)]$, which is zero except for $n = 1$ in which case $\alpha_1 = \sqrt{2}/2$, and
$$\gamma_n = \int_0^1 (-x)\sqrt{2}\sin[(2n-1)\pi x/2]dx \quad [\text{Eq.(35) with}$$
$F(x,t) = -x]$. From Problem 2 we have
$$\gamma_n = -4\sqrt{2}(-1)^{n+1}/(2n-1)^2\pi^2 = -c_n, \quad n = 1,2,\ldots \ . \ \text{Setting}$$
$\gamma_n = -c_n$ in Eq.(42) we then have
$$b_1 = \frac{\sqrt{2}}{2}e^{-\pi^2 t/4} - c_1\int_0^t e^{-\pi^2(t-s)/4}ds$$
$$= \frac{\sqrt{2}}{2}e^{-\pi^2 t/4} - \frac{4c_1}{\pi^2}e^{-\pi^2(t-s)/4}\Big|_0^t$$
$$= \frac{\sqrt{2}}{2}e^{-\pi^2 t/4} - \frac{4c_1}{\pi^2} + \frac{4c_1}{\pi^2}e^{-\pi^2 t/4} \quad \text{and}$$
similarly $b_n = -c_n\int_0^t e^{-\lambda_n(t-s)}ds = -(c_n/\lambda_n)e^{-\lambda_n(t-s)}\Big|_0^t$
$$= -(c_n/\lambda_n)(1-e^{-\lambda_n t}), \quad \text{where } \lambda_n = (2n-1)^2\pi^2/4,$$
$n = 2,3,\ldots \ .$ Substituting these values for b_n along with $\phi_n = \sqrt{2}\sin[(2n-1)\pi x/2]$ into the series for $u(x,t)$ yields the solution to the given problem.

22. In this case $\alpha_n = 0$ for all n and γ_n is given by

$$\gamma_n = \int_0^1 e^{-t}(1-x)\sqrt{2}\sin[(2n-1)\pi x/2]dx$$

$$= e^{-t}\int_0^1 (1-x)\sqrt{2}\sin[(2n-1)\pi x/2]dx.$$ This last integral

can be written as the sum of two integrals, each of which has been evaluated in either Problem 6 or 7 of Section 11.3. Letting c_n denote the value obtained, we then have

$$\gamma_n = c_n e^{-t} \text{ and thus } b_n = c_n\int_0^t e^{-\lambda_n(t-s)}e^{-s}ds =$$

$$c_n e^{-\lambda_n t}\int_0^t e^{(\lambda_n-1)s}ds = [c_n/(\lambda_n-1)](e^{-t}-e^{-\lambda_n t}), \text{ where}$$

$\lambda_n = (2n-1)^2\pi^2/4$. Substituting these values into Eq.(30) yields the desired solution.

24. Using the approach of Problem 23 we find that v(x) satisfies $v'' = 2$, $v(0) = 1$, $v(1) = 0$. Thus $v(x) = x^2 + c_1 x + c_2$ and the B.C. yield $v(0) = c_2 = 1$ and $v(1) = 1 + c_1 + 1 = 0$ or $c_1 = -2$. Hence $v(x) = x^2 - 2x + 1$ and $w(x,t) = u(x,t) - v(x)$ where, from Problem 23, we have $w_t = w_{xx}$, $w(0,t) = 0$, $w(1,t) = 0$ and $w(x,0) = x^2 - 2x + 2 - v(x) = 1$. This last problem can be solved by methods of this section or by methods of Chapter 10. Using the approach of this section we have

$$w(x,t) = \sum_{n=1}^{\infty} b_n(t)\phi_n(x) \text{ where } \phi_n(x) = \sqrt{2}\sin n\pi x$$

[which are the normalized eigenfunctions of the associated eigenvalue problem $y'' + \lambda y = 0$, $y(0) = 0$, $y(1) = 0$] and the b_n are given by Eq.(42). Since the P.D.E. for $w(x,t)$ is homogeneous Eq.(42) reduces to $b_n = \alpha_n e^{-\lambda_n t}$ ($\lambda_n = n^2\pi^2$ from the above eigenvalue problem), where

$$\alpha_n = \int_0^1 1\cdot\sqrt{2}\sin n\pi x dx = \sqrt{2}[1(-1)^n]/n\pi.$$ Thus

$$u(x,t) = x^2-2x + 1 + \sum_{n=1}^{\infty} \frac{\sqrt{2}[1-(-1)^n]}{n\pi} e^{-n^2\pi^2 t}\sqrt{2}\sin n\pi x,$$

which simplifies to the desired solution.

28a. Since $y_c = c_1 + c_2 x$, we assume that $Y(x) = u_1(x) + xu_2(x)$. Then $Y' = u_2$ since we require $u_1' + xu_2' = 0$. Differentiating again yields $Y'' = u_2'$ and thus $u_2' = -f(x)$ by substitution into the D.E. Hence

$u_2(x) = -\int_0^x f(s)\,ds,\ u_1' = xf(x),$ and $u_1(x) = \int_0^x sf(s)\,ds.$

Therefore $Y = \int_0^x sf(s)\,ds - x\int_0^x f(s)\,ds = -\int_0^x (x-s)f(s)\,ds$ and

$\phi(x) = c_1 + c_2 x - \int_0^x (x-s)f(s)\,ds.$

28c. From parts a and b we have

$$\phi(x) = x\int_0^1 (1-s)f(s)\,ds - \int_0^x (x-s)f(s)\,ds$$

$$= \int_0^x x(1-s)f(s)\,ds + \int_x^1 x(1-s)f(s)\,ds - \int_0^x (x-s)f(s)\,ds$$

$$= \int_0^x (x-xs-x+s)f(s)\,ds + \int_x^1 x(1-s)f(s)\,ds$$

$$= \int_0^x s(1-x)f(s)\,ds + \int_x^1 x(1-s)f(s)\,ds.$$

30b. In this case $y_1(x) = \sin x$ and $y_2(x) = \sin(1-x)$ [assume
$y_2(x) = c_1\cos x + c_2\sin x$, let $x = 1$, solve for c_2 in terms
of c_1 using $y(1) = 0$ and then let $c_1 = \sin 1$]. Using
these functions for y_1 and y_2 we find $W(y_1,y_2) = -\sin 1$
and thus $G(x,s) = -\sin s\,\sin(1-x)/(-\sin 1)$, since $p(x) = 1$,
for $0 \le s \le x$. Interchanging the x and s verifies $G(x,s)$
for $x \le s \le 1$.

30c. Since $W(y_1,y_2)(x) = y_1(x)y_2'(x) - y_2(x)y_1'(x)$ we find that

$$[p(x)W(y_1,y_2)(x)]' = p'(x)[y_1(x)y_2'(x) - y_2(x)y_1'(x)]$$

$$+ p(x)[y_1'(x)y_2'(x) + y_1(x)y_2''(x) - y_2'(x)y_1'(x) - y_2(x)y_1''(x)]$$

$$= y_1[py_2']' - y_2[py_1']' = y_1[q(x)y_2] - y_2[q(x)y_1] = 0.$$

30d. Let $c = p(x)W(y_1,y_2)(x)$. If $0 \le s \le x$, then
$G(x,s) = -y_1(s)y_2(x)/c$. Since the first argument in
$G(s,x)$ is less than the second argument, the bottom
expression of formula (iv) must be used to determine
$G(s,x)$. Thus, $G(s,x) = -y_1(s)y_2(x)/c$. A similar argument
holds if $x \le s \le 1$.

33. In general $y(x) = c_1\cos x + c_2\sin x$. For $y'(0) = 0$ we must
choose $c_2 = 0$ and thus $y_1(x) = \cos x$. For $y(1) = 0$ we have
$c_1\cos 1 + c_2\sin 1 = 0$, which yields $c_2 = -c_1(\cos 1)/\sin 1$ and

thus $y_2(x) = c_1\cos x - c_1(\cos 1)\sin x/\sin 1$

$\qquad = c_1(\sin 1\cos x - \cos 1\sin x)/\sin 1$

$\qquad = \sin(1-x)$ [by setting $c_1 = \sin 1$].

Furthermore, $W(y_1, y_2) = -\cos 1$ and thus

$$G(x,s) = \begin{cases} \dfrac{\cos s \sin(1-x)}{\cos 1} & 0 \le s \le x \\[2ex] \dfrac{\cos x \sin(1-s)}{\cos 1} & x \le s \le 1 \end{cases},$$

and hence $\phi(x) = \displaystyle\int_0^x [\cos s \sin(1-x) f(s)/\cos 1]ds +$

$\displaystyle\int_x^1 [\cos x \sin(1-s) f(s)/\cos 1]ds$ is the solution of the given

B.V.P.

Section 11.5, Page 664

2a. From Eq.(9), the general solution of the D.E. is
$y = c_1 J_0(\sqrt{\lambda}\, x) + c_2 Y_0(\sqrt{\lambda}\, x)$. The B.C. at $x = 0$ requires
that $c_2 = 0$, and the B.C. at $x = 1$ requires
$c_1 \sqrt{\lambda}\, J_0'(\sqrt{\lambda}) = 0$. For $\lambda = 0$ we have $\phi_0(x) = J_0(0) = 1$
and if λ_n is the nth positive root of $J_0'(\sqrt{\lambda}) = 0$ then
$\phi_n(x) = J_0(\sqrt{\lambda_n}\, x)$. Note that for $\lambda = 0$ the D.E. becomes
$(xy')' = 0$, which has the general solution $y = c_1\ln x + c_2$.
To satisfy the bounded conditions at $x = 0$ we must choose
$c_1 = 0$, thus obtaining the same solution as above.

2b. For $n \ne 0$, set $y = J_0(\sqrt{\lambda_n}\, x)$ in the D.E. and integrate
from 0 to 1 to obtain $-\displaystyle\int_0^1 (xJ_0')'dx = \lambda_n \int_0^1 xJ_0(\sqrt{\lambda_n}\, x)dx$.
Integrating the left side of this equation yields
$\displaystyle\int_0^1 (xJ_0')'dx = xJ_0'(\sqrt{\lambda_n}\, x)\Big|_0^1 = J_0'(\sqrt{\lambda_n}) - 0 = 0$ since the λ_n
are eigenvalues from part a. Thus $\displaystyle\int_0^1 xJ_0(\sqrt{\lambda_n}\, x)dx = 0$.
For other n and m, we let $L[y] = -(xy')'$. Then
$L[J_0(\sqrt{\lambda_n}\, x)] = \lambda_n xJ_0(\sqrt{\lambda_n}\, x)$ and
$L[J_0(\sqrt{\lambda_m}\, x)] = \lambda_m xJ_0(\sqrt{\lambda_m}\, x)$. Multiply the first equation
by $J_0(\sqrt{\lambda_m}\, x)$, the second by $J_0(\sqrt{\lambda_n}\, x)$, subtract the
second from the first, and integrate from 0 to 1 to
obtain

$$\int_0^1 \{J_0(\sqrt{\lambda_m}\,x)\,L[J_0(\sqrt{\lambda_n}\,x)] - J_0(\sqrt{\lambda_n}\,x)\,L[J_0(\sqrt{\lambda_m}\,x)]\}\,dx =$$

$(\lambda_n - \lambda_m)\int_0^1 x J_0(\sqrt{\lambda_n}\,x)\,J_0(\sqrt{\lambda_m}\,x)\,dx$. Again the left side is zero after each term is integrated by parts once, as was done above. If $\lambda_n \neq \lambda_m$, the result follows with

$$\phi_n(x) = J_0(\sqrt{\lambda_n}\,x).$$

2c. We assume that $y = b_0 + \sum_{n=1}^{\infty} b_n J_0(\sqrt{\lambda_n}\,x)$. Since

$-[x J_0'(\sqrt{\lambda_n}\,x)]' = \lambda_n x J_0(\sqrt{\lambda_n}\,x)$, $n = 0,1,\ldots$, we find that

$-(xy')' = x\sum_{n=1}^{\infty}\lambda_n b_n J_0(\sqrt{\lambda_n}\,x)$ [note that $\lambda_0 = 0$ and hence b_0

is missing on the right]. Now assume

$f(x)/x = c_0 + \sum_{n=1}^{\infty} c_n J_0(\sqrt{\lambda_n}\,x)$. Multiplying both sides by

$x J_0(\sqrt{\lambda_m}\,x)$, integrating from 0 to 1 and using the orthogonality relations of part b, we find

$c_n = \int_0^1 f(x)\,J_0(\sqrt{\lambda_n}\,x)\,dx / \int_0^1 x J_0^2(\sqrt{\lambda_n}\,x)\,dx$, $n = 0,1,2,\ldots$.

[Note that $c_0 = 2\int_0^1 f(x)\,dx$ since the denominator can be

integrated.] Substituting the series for y and $f(x)/x$ into the D.E., using the above result for $-(xy')'$, and simplifying we find that

$(\mu b_0 + c_0) + \sum_{n=1}^{\infty} [c_n - b_n(\lambda_n - \mu)]J_0(\sqrt{\lambda_n}\,x) = 0$. Thus

$b_0 = -c_0/\mu$ and $b_n = c_n/(\lambda_n - \mu)$, $n = 1,2,\ldots$, where $\sqrt{\lambda_n}$

are obtained from $J_0'(\sqrt{\lambda_n}) = 0$.

4a. Let $L[y] = -[(1-x^2)y']'$. Then $L[\phi_n] = \lambda_n\phi_n$ and

$L[\phi_m] = \lambda_m\phi_m$. Multiply the first equation by ϕ_m, the second by ϕ_n, subtract the second from the first, and integrate from 0 to 1 to obtain

$\int_0^1 (\phi_m L[\phi_n] - \phi_n L[\phi_m])\,dx = (\lambda_n - \lambda_m)\int_0^1 \phi_n\phi_m\,dx$. The integral

on the left side can be shown to be 0 by integrating each term once by parts. Since $\lambda_n \neq \lambda_m$ if $m \neq n$, the result follows. Note that the result may also be written as

$\int_0^1 P_{2m-1}(x)\,P_{2n-1}(x)\,dx = 0$, $m \neq n$.

4b. First let $f(x) = \sum\limits_{n=1}^{\infty} c_n \phi_n(x)$, multiply both sides by $\phi_m(x)$,
and integrate term by term from $x = 0$ to $x = 1$. The
orthogonality condition yields

$c_n = \int_0^1 f(x)\phi_n(x)\,dx / \int_0^1 \phi_n^2(x)\,dx$, $n = 1,2,\ldots$ where it is
understood that $\phi_n(x) = P_{2n-1}(x)$. Now assume

$y = \sum\limits_{n=1}^{\infty} b_n \phi_n(x)$. As in Problem 2 and in the text

$-[(1-x^2)y']' = \sum\limits_{n=1}^{\infty} \lambda_n b_n \phi_n$ since the ϕ_n are eigenfunctions.

Thus substitution of the series for y and f into the D.E.
and simplification yields $\sum\limits_{n=1}^{\infty} [b_n(\lambda_n - \mu) - c_n]\phi_n(x) = 0$.

Hence $b_n = c_n/(\lambda_n - \mu)$, $n = 1,2,\ldots$ and the desired
solution is obtained [after setting $\phi_n(x) = P_{2n-1}(x)$].

Section 11.6, Page 669

1a. Since $u(x,0) = 0$ we have $Y(0) = 0$. However, since the
other two boundaries are given by $y = 2x$ and $y = 2(x-2)$
we cannot separate x and y dependence and thus neither X
nor Y satisfy homogeneous B.C. at both end points.

1b. The line $y = 2x$ is transformed into $\xi = 0$ and $y = 2(x-2)$
is transformed into $\xi = 2$, so the parallelogram is
transformed into a square of side 2.

2. This problem is very similar to the example worked in the
text. The fundamental solutions satisfying the
P.D.E.(3), the B.C. $u(1,t) = 0$, $t \geq 0$ and the finiteness
condition are given by Eqs.(15) and (16). Thus assume
$u(r,t)$ is of the form given by Eq.(17). The I.C. require
that $u(r,0) = \sum\limits_{n=1}^{\infty} c_n J_0(\lambda_n r) = 0$ and

$u_t(r,0) = \sum\limits_{n=1}^{\infty} \lambda_n a k_n J_0(\lambda_n r) = g(r)$. From Eq.(26) of Section
11.5 we obtain $c_n = 0$ and
$\lambda_n k_n a = \int_0^1 rg(r)J_0(\lambda_n r)\,dr / \int_0^1 rJ_0^2(\lambda_n r)\,dr$, $n = 1,2,\ldots$.

4.   This problem is the same as Problem 24 of Section 10.6. The periodicity condition requires that μ of that problem be an integer and thus substituting $\mu^2 = n^2$ into the previous results yields the given equations.

5a. Substituting $u(r,\theta,z) = R(r)\Theta(\theta)Z(z)$ into Laplace's equation yields $R''\Theta Z + R'\Theta Z/r + R\Theta''Z/r^2 + R\Theta Z'' = 0$ or equivalently $R''/R + R'/rR + \Theta''/r^2\Theta = -Z''/Z = \sigma$. In order to satisfy arbitrary B.C. it can be shown that σ must be negative, so assume $\sigma = -\lambda^2$, and thus $Z'' - \lambda^2 Z = 0$ and, after some algebra, it follows that $r^2 R''/R + rR'/R + \lambda^2 r^2 = -\Theta''/\Theta = \alpha$. The periodicity condition $\Theta(0) = \Theta(2\pi)$ requires that $\sqrt{\alpha}$ be an integer n so $\alpha = n^2$. Thus $r^2 R'' + rR' + (\lambda^2 r^2 - n^2)R = 0$, $\Theta'' + n^2\Theta = 0$, and $Z'' - \lambda^2 Z = 0$.

5b. If $u(r,\theta,z)$ is independent of θ, then the $\Theta''/r^2\Theta$ term does not appear in the second equation of part a and thus $R''/R + R'/rR = -Z''/Z = -\lambda^2$, from which the desired result follows.

6.   Assuming that $u(r,z) = R(r)Z(z)$ it follows from Problem 5 that $R = c_1 J_0(\lambda r) + c_2 Y_0(\lambda r)$ and $Z = k_1 e^{-\lambda z} + k_2 e^{\lambda z}$. Since $u(r,z)$ is bounded as $r \to 0$ and $z \to \infty$ we require that $c_2 = 0$, $k_2 = 0$. The B.C. $u(1,z) = 0$ requires that $J_0(\lambda) = 0$ leading to an infinite set of discrete positive eigenvalues $\lambda_1, \lambda_2, \ldots \lambda_n \ldots$. The fundamental solutions of the problem are then $u_n(r,z) = J_0(\lambda_n r)e^{-\lambda_n z}$, $n = 1,2,\ldots$. Thus assume $u(r,z) = \sum_{n=1}^{\infty} c_n J_0(\lambda_n r)e^{-\lambda_n z}$. The B.C. $u(r,0) = f(r)$, $0 \le r \le 1$ requires that $u(r,0) = \sum_{n=1}^{\infty} c_n J_0(\lambda_n r) = f(r)$ so $c_n = \int_0^1 rf(r)J_0(\lambda_n r)dr / \int_0^1 rJ_0^2(\lambda_n r)dr$, $n = 1,2,\ldots$.

7b. Again Θ periodic of period 2π implies $\lambda^2 = n^2$. Thus the solutions to the D.E. are $R(r) = c_1 J_n(kr) + c_2 Y_n(kr)$ and $\Theta(\theta) = d_1\cos n\theta + d_2\sin n\theta$, $n = 0,1,2\ldots$. For the solution to remain bounded, $c_2 = 0$ and thus $v(r,\theta) = (1/2)c_0 J(kr) + \sum_{m=1}^{\infty} J_m(kr)(b_m\sin m\theta + c_m\cos m\theta)$. Hence

$v(c,\theta)$ is then a Fourier Series of period 2π and the coefficients are found by Eqs.(13) and (14) and Problem 27 of Section 10.2.

9a. Substituting $u(\rho,\theta,\phi) = P(\rho)\Theta(\theta)\Phi(\phi)$ into Laplace's equation leads to
$\rho^2 P''/P + 2\rho P'/P = -(\csc^2\phi)\Theta''/\Theta - \Phi''/\Phi - (\cot\phi)\Phi'/\Phi = \sigma.$
In order to satisfy arbitrary B.C. it can be shown that σ must be positive, so assume $\sigma = \mu^2$.
Thus $\rho^2 P'' + 2\rho P' - \mu^2 P = 0$. Then we have
$(\sin^2\phi)\Phi''/\Phi + (\sin\phi\cos\phi)\Phi'/\Phi + \mu^2\sin^2\phi = -\Theta''/\Theta = \alpha.$
The periodicity condition $\Theta(0) = \Theta(2\pi)$ requires that $\sqrt{\alpha}$ be an integer λ so $\alpha = \lambda^2$. Hence $\Theta'' + \lambda^2\Theta = 0$ and
$(\sin^2\phi)\Phi'' + (\sin\phi\cos\phi)\Phi' + (\mu^2\sin^2\phi - \lambda^2)\Phi = 0.$

10. Since u is independent of θ, only the first and third of the Eqs. in 9a hold. The general solution to the Euler equation is

$P = c_1\rho^{r_1} + c_2\rho^{r_2}$ where $r_1 = (-1+\sqrt{1+4\mu^2})/2 > 0$ and
$r_2 = (-1-\sqrt{1+4\mu^2})/2 < 0$. Since we want u to be bounded as $\rho \to 0$, we set $c_2 = 0$. As found in Problem 22 of Section 5.3, the solutions of Legendre's equation, Problem 9c, are either singular at 1, at -1, or at both unless $\mu^2 = n(n+1)$, where n is an integer. In this case, one of the two linearly independent solutions is a polynomial denoted by P_n (Problems 23 and 24 of Section 5.3). Since $r_1 = (-1 + \sqrt{1+4n(n+1)})/2 = n$, the fundamental solutions of this problem satisfying the finiteness condition are $u_n(\rho,\phi) = \rho^n P_n(s) = \rho^n P_n(\cos\phi)$, $n = 1,2,\ldots$. It can be shown that an arbitrary piecewise continuous function on $[-1,1]$ can be expressed as a linear combination of Legendre polynomials. Hence we assume that

$u(\rho,\phi) = \displaystyle\sum_{n=1}^{\infty} c_n\rho^n P_n(\cos\phi)$. The B.C. $u(1,\phi) = f(\phi)$ requires

that $u(1,\phi) = \displaystyle\sum_{n=1}^{\infty} c_n P_n(\cos\phi) = f(\phi)$, $0 \le \phi \le \pi$. From

Problem 28 of Section 5.3 we know that $P_n(x)$ are orthogonal. However here we have $P_n(\cos\phi)$ and thus we must rewrite the equation in Problem 9b to find
$-[(\sin\phi)\Phi']' = \mu^2(\sin\phi)\Phi.$ Thus $P_n(\cos\phi)$ and $P_m(\cos\phi)$ are

orthogonal with weight function $\sin\phi$. Thus we must multiply the series expansion for $f(\phi)$ by $\sin\phi P_m(\cos\phi)$ and integrate from 0 to π to obtain $c_m = \int_0^\pi f(\phi)\sin\phi P_m(\cos\phi)\,d\phi / \int_0^\pi \sin\phi P_m^2(\cos\phi)\,d\phi$. To obtain the answer as given in the text let $s = \cos\phi$.

Section 11.7, Page 678

1a. Write $S_n(x)$ as $n\sqrt{x}/e^{nx^2/2}$ and use L'Hopitals Rule.

3. Since $f(x)$ is defined only on the open interval we see that $f(x)$ can get as close to 1 as desired, but $f(x)$ is never equal to 1. Thus the least upper bound is 1, but there is no maximum value.

7. Expanding the integrand we get
$$R_n = \int_0^1 r(x)[f(x) - S_n(x)]^2 dx = \int_0^1 r(x)f^2(x)\,dx$$
$$-2\sum_{i=1}^n c_i\int_0^1 r(x)f(x)\phi_i(x)\,dx + \sum_{i=1}^n \sum_{j=1}^n c_i c_j\int_0^1 r(x)\phi_i(x)\phi_j(x)\,dx,$$
where the last term is obtained by calculating $S_n^2(x)$. Using Eqs.(1) and (9) this becomes
$$R_n = \int_0^1 r(x)f^2(x)\,dx - 2\sum_{i=1}^n c_i a_i + \sum_{i=1}^n c_i^2$$
$$= \int_0^1 r(x)f^2(x)\,dx - \sum_{i=1}^n a_i^2 + \sum_{i=1}^n (c_i - a_i)^2, \text{ by completing}$$
the square. Since all terms involve a real quantity squared (and $r(x) > 0$) we may conclude R_n is minimized by choosing $c_i = a_i$. This can also be shown by calculating $\partial R_n/\partial c_i = 2(c_i - a_i)$ and setting equal to zero.

9b. From part a we have $f_0(x) = 1$ and thus $f_1(x) = c_1 + c_2 x$ must satisfy $(f_0, f_1) = \int_0^1 (c_1 + c_2 x)\,dx = 0$ and $(f_1, f_1) = \int_0^1 (c_1 + c_2 x)^2 dx = 1$. Evaluating the integrals yields $c_1 + c_2/2 = 0$ and $c_1^2 + c_1 c_2 + c_2^2/3 = 1$, which have the solution $c_1 = \sqrt{3}$, $c_2 = -2\sqrt{3}$ and thus $f_1(x) = \sqrt{3}(1-2x)$.

9c. $f_2(x) = c_1 + c_2x + c_3x^2$ must satisfy $(f_0, f_2) = 0$,
     $(f_1, f_2) = 0$ and $(f_2, f_2) = 1$.

9d. For $g_2(x) = c_1 + c_2x + c_3x^2$ we have $(g_0, g_2) = 0$ and
     $(g_1, g_2) = 0$, which yield the same ratio of coefficients
     as found in 9c. Thus $g_2(x) = cf_2(x)$, where c may now be
     found from $g_2(1) = 1$.

10. This problem follows the pattern of Problem 9 except now
     the limits on the orthogonality integral are from -1 to
     1. That is $(P_i, P_j) = \int_{-1}^{1} P_i(x) P_j(x)\, dx = 0$, $i \neq j$. For
     $i = 0$ we have $P_0(x) = 1$ and for $i = 0$ and $j = 1$ we

     have $(P_0, P_1) = \int_{-1}^{1} (c_1 + c_2x)\, dx = \left. (c_1x + c_2x^2/2) \right|_{-1}^{1} = 2c_1 = 0$ and

     thus $P(1) = 1$ yields $P_1(x) = x$. The others follow in a
     similar fashion.

11a. This part has essentially been worked in Problem 7 by
     setting $c_i = a_i$.

11b. Eq. (6) shows that $R_n \geq 0$ since $r(x) \geq 0$ and thus

     $\int_{0}^{1} r(x) f^2(x)\, dx - \sum_{i=1}^{n} a_i^2 \geq 0$. The result follows.

11c. Since f is square integrable, $\int_{0}^{1} r(x) f^2(x)\, dx = M < \infty$ and
     therefore the monotone increasing sequence of partial

     sums $T_n = \sum_{i=1}^{n} a_i^2$ is bounded above. Thus $\lim_{n \to \infty} T_n$ exists,

     which proves the convergence of the given sum.

11e. By definition if $\sum_{i=1}^{\infty} a_i \phi_i(x)$ converges to $f(x)$ in the

     mean, then $R_n \to 0$ as $n \to \infty$. Hence $\int_{0}^{1} r(x) f^2(x)\, dx = \sum_{i=1}^{\infty} a_i^2$.

     Conversely, if $\int_{0}^{1} r(x) f^2(x)\, dx = \sum_{i=1}^{\infty} a_i^2$, $\lim_{n \to \infty} R_n = 0$ and

     $\sum_{i=1}^{\infty} a_i \phi_i(x)$ converges to $f(x)$ in the mean.

12. Bessel's inequality implies that $\sum\limits_{i=1}^{\infty} a_i^2$ converges and thus

the nth term $a_n \to 0$ as $n \to \infty$.

14. If the series were the eigenfunction series for a square

integrable function, the series $\sum\limits_{i=1}^{\infty} a_i^2$ would have to

converge. But $a_0 = 1$, $a_1 = 1/\sqrt{2}, \ldots, a_n = 1/\sqrt{n}, \ldots,$

and $\sum\limits_{n=1}^{\infty} a_n^2 = \sum\limits_{n=1}^{\infty} 1/n$ is the well-known harmonic series which

does not converge.